ABOUT THE AUTHOR

Brian Fletcher started his career as a Research Physicist specializing in accurate measurements of extremely low pressures. This was followed by many years teaching Mathematics up to University Entrance standard. The final five years of his career were spent teaching Information Technology to adults.

Now retired, he lives with his wife in a small village in Wiltshire spending some of his time as a volunteer crew member and helmsman for a nearby canal charity and is also a Governor of the local primary school.

His previous publication *'Anyone Can Do Arithmetic'* indicates the standard of Arithmetic required to fully appreciate this book. It can be ordered from most bookshops and from the publisher. It is also available in many public libraries.

ANYONE CAN DO ALGEBRA

BRIAN FLETCHER

Matador
9 Priory Business Park,
Wistow Road, Kibworth Beauchamp,
Leicestershire. LE8 0RX
Tel: (+44) 116 279 2299
Fax: (+44) 116 279 2277
Email: books@troubador.co.uk
Web: www.troubador.co.uk/matador

ISBN 978 1784621 261

British Library Cataloguing in Publication Data.
A catalogue record for this book is available from the British Library.

Printed and bound in the UK by TJ International, Padstow, Cornwall
Typeset in 11pt Garamond by Troubador Publishing Ltd, Leicester, UK

Matador is an imprint of Troubador Publishing Ltd

MIX
Paper from
responsible sources
FSC
www.fsc.org   FSC® C013056

ACKNOWLEDGEMENTS

Once again I am indebted to my son, Graham, for many helpful suggestions and his meticulous proof reading. I am also grateful to my friends who have read *'Arithmetic'* and have encouraged this publication. My wife, Sheila, has again been patient with my neglect of domestic duties while writing this book.

Brian Fletcher 2014

CONTENTS

INTRODUCTION

In order to understand Algebra the reader will need Arithmetic skills to the standard described in the author's previous book *'Anyone Can Do Arithmetic'*. (This book will be frequently referred to, and for ease of typing will be called simply *'Arithmetic'*.)

Many people are convinced that they cannot do Algebra and pose the question 'why should I need to know anything about it?'.

One answer, of course, is that you cannot continue to make much more progress with Mathematics without a fairly good knowledge of Algebra. This answer will only satisfy a minority.

There is, however, a much more practical reason for needing some knowledge of Algebra. As I write this, the UK has recently emerged from one recession and is now in another one, global finances are in a mess and the level of unemployment is very high. There are many jobs where you may think Mathematical skills would not be needed, but a large proportion of employers would prefer applicants who have obtained a reasonable GCSE grade in Mathematics as it demonstrates the ability for logical thought and problem solving.

Algebra – probably the most feared word in the classroom. Unfortunately Algebra is usually introduced to pupils as a new topic. In reality, basic Algebra needs no more skills than those used in Arithmetic. In Arithmetic we are presented with some numbers and are told to combine them in various ways to produce a numerical answer. Algebra can be thought of as the reverse of this. The calculation includes an unknown number and we are given the answer. The Arithmetic used in the calculation has to be undone leading us to the unknown number.

One of the simplest ways to see Algebra at work is in the age old childhood puzzle; "I am thinking of a number. I then add three and multiply this answer by five. The result is thirty five. What number did I start with?" I am sure most readers will have already worked out that the number was four. Just to state this problem uses a lot of words and what causes difficulty for many is the use of a symbol for the unknown quantity – in particular the dreaded letter 'x'. Of course, any letter would do, or even a short word. However, letters from the end of the alphabet (usually x, y, and z) to represent unknown quantities have been in use since their introduction in the 17th century by a French mathematician called René Descartes (1596 – 1650), so it is a bit late to change this now.

Using the symbol x for the unknown number we can now state the above problem very efficiently using the equation $5(x + 3) = 35$. This shows very clearly that the unknown number first has three added to it and this answer is multiplied by five to produce the final answer of thirty five. (As explained in *'Arithmetic'* it is normal to omit the multiplication sign between a number and a bracket.)

It is worth emphasizing that the equation shown in the last paragraph obeys the primary rule of all equations – the quantity to the left of the equals sign has exactly the same value as the quantity to the right. In this case there is only one value that can replace x to make the equation true.

Using a symbol to stand for an unknown number is just one way in which letters are used in equations. Another important use of letters is to use a letter to stand for a variable quantity. This is best explained by use of a simple example.

Most people are familiar with the fact that average speed can be worked out by dividing the distance travelled by the time taken. Instead of this lengthy sentence we can use the letter S in place of average speed, D instead of distance travelled and T instead of time taken. Now we can write a simple equation linking these three quantities, namely;

$$S = D \div T.$$

The letters S, D and T are called variables as we can choose any values for two of them and then use the equation to work out the value of the third. You may be more familiar with this when the equation is called a formula. It contains the instructions for working out one of these quantities when the other two have values assigned to them.

In this book you will see many examples of the use of variables. The idea of using variables is a very powerful tool. It enables a formula to be constructed which will give you the rules for solving a whole range of equations.

Equations and their solutions are at the heart of Algebra. The simplest equations are those that only involve multiples of x and numbers. For reasons which will become apparent in Chapter 7 these are called linear equations and their solution is fairly simple. When multiples of x^2 are included we have what is called a quadratic equation. You may recall that x^2 is said as 'x squared' and the prefix quad in quadratic is to do with squares, as in quadrangle. Before solving these equations you will be introduced to the idea of a square root which is an essential first step. Equations of this type are extremely important and lead to further developments in Mathematics.

INTRODUCTION

Readers who have read *'Arithmetic'* will find that this book proceeds at a slightly faster pace. The reason for this is that all the groundwork was done in *'Arithmetic'*, and most of this book is devoted to understanding the use of letters to denote solutions and variables.

The case for understanding Algebra does get debated by our Governments. During a debate on the School Curriculum on 26 June 2003 Mr. McWalter MP said a quadratic equation is a door to a room full of the unparalleled riches of human intellectual achievement.

This is probably a little over the top but Mr. McWalter obviously believes that understanding Algebra is very important.

Some advice to the reader is not to be put off by your fear of Algebra. It is lack of understanding that should be feared. Do not be worried about trying a new way of solving problems. If your way works and the mathematical reasoning is sound then you have learned something. Equally, if your method fails and you can see why it failed you have also learned something.

As in *'Arithmetic'* the best way of reading this book for the first time is in the order of the Chapters. Please read on and allow me to prove that 'Anyone Can Do Algebra'.

CHAPTER ONE

ARITHMETIC WITH ALGEBRA

Arithmetic and Algebra together.
Does this mean I have to be clever?
No, no need to feel queasy;
Read on and it will become easy.

It is first necessary to recognize Arithmetic that we cannot do with Algebra before embarking on what is possible.

A sum like $2x + 5$ cannot be simplified as there is no way we can multiply a number by 2 and then add 5 until we know what the number was in the first place, so we have to be content with this as an answer. In the same way an expression such as $3x^2 - 9x + 4$ cannot be simplified.

Adding and subtracting with Algebra is normally just a process of tidying. We can obviously add and subtract numbers using the normal rules of Arithmetic. Multiples of x such as $7x - 2x$ can be simplified to make $5x$ just as we would simplify £7 − £2 to get an answer of £5.

Terms involving x^2 can be combined, e.g. $3x^2 + 6x^2 = 9x^2$, similarly with terms involving x^3, x^4, etc.

Using the ideas of adding fractions explained in *'Arithmetic'* we can complete the following types of additions and subtractions:

$$\frac{2}{x} + \frac{3}{x} = \frac{5}{x}$$

$$\frac{6}{x^2} - \frac{2}{x^2} = \frac{4}{x^2}$$

At this stage a reminder of columns headings in Arithmetic is worthwhile. As tenths, hundredths and thousandths can be written as $1/10$, $1/10^2$ and $1/10^3$ we can show the column headings as:

10^3	10^2	10	Units	$\dfrac{1}{10}$	$\dfrac{1}{10^2}$	$\dfrac{1}{10^3}$

This compares very nicely with the column headings we can achieve with Algebra, namely:

x^3	x^2	x	Numbers	$\dfrac{1}{x}$	$\dfrac{1}{x^2}$	$\dfrac{1}{x^3}$

Most addition and subtraction with Algebra is concerned with terms involving x^2, x and numbers. All the terms in x^2 can be combined using the rules of Arithmetic, as can the terms in x and the number terms. So the complicated looking expression $8x + 3x^2 - 5x + 17 + x - x^2 + 5$ can be simplified to become $2x^2 + 4x + 22$.

It is worth noting here that, unless there is a good reason to do otherwise, it is usual to start the answer with the highest powers of x and finish with the numbers thus following the order of the table above.

Any more difficult examples of addition and subtraction will be dealt with as they arise.

As in 'Arithmetic' each topic is introduced with simple examples and multiplication is no exception. When multiplying using numbers there are a few answers that need to be learned and multiplying with Algebra is no different. A few examples follow:

$$7 \times x = 7x$$
$$x \times x = x^2$$
$$3 \times x \times x = 3x^2$$
$$x^2 \times x = x^3$$
$$5x \times x = 5x^2$$
$$(5x)^2 = 25x^2$$

2

The most important type of multiplying with Algebra is when using expressions involving two terms such as $3x + 2$. This is called a binomial. A simple multiplication using a binomial would be $4(3x + 2)$. As shown in Chapter 10 of 'Arithmetic' this is worked out as $4 \times 3x + 4 \times 2 = 12x + 8$.

Similarly, the answer to $7(2x - 3)$ is $14x - 21$.

We can now show how useful the idea of a variable can be by constructing a simple formula for working out all calculations of this type. If a, b and c stand for any numbers of our choice we can write the formulae (this is the plural of formula):

$$a(bx + c) = abx + ac$$

and:

$$a(bx - c) = abx - ac.$$

Using the plus or minus sign these can usefully be combined into one:

$$a(bx \pm c) = abx \pm ac.$$

We are now in a position to find out how to multiply two binomials. This will be done by comparison with numerical calculations. If the calculation is 14×23 we know from the explanation in 'Arithmetic' that it can be written as:

$$(10 + 4)(20 + 3)$$

and the four intermediate answers to be worked out are:

$$10 \times 20$$
$$10 \times 3$$
$$4 \times 20$$
$$4 \times 3$$

The rules of Arithmetic must apply to all numbers even when the value is not yet known. So, by comparing $(2x + 1)(x + 3)$ with the previous paragraph we can write the four answers:

$$2x \times x = 2x^2$$
$$2x \times 3 = 6x$$
$$1 \times x = x$$
$$1 \times 3 = 3.$$

$6x + x = 7x$, so the answer can be written as:

$$(2x + 1)(x + 3) = 2x^2 + 7x + 3$$

There is a word that is used to remind you of the four calculations that are done when two binomials are multiplied. It is FOIL; the F standing for Firsts, the O for outsides, the I for insides and the L for Lasts.

So in the sum $(3x + 2)(2x - 6)$, F comes from $3x \times 2x = 6x^2$, O is from $3x \times (-6) = (-18x)$, I is $2 \times 2x = 4x$ and L is $2 \times (-6) = (-12)$. Combining these four terms gives $6x^2 + (-18x) + 4x - 12 = 6x^2 - 14x - 12$.

At this stage it will be useful to construct a formula which will allow easy multiplication of any pair of binomials. To do this four numbers are needed, so a, b, c and d are used to represent any numbers of our choice. The two binomials can then be written as $(ax + b)$ and $(cx + d)$ and following the idea of the last paragraph we can write:

$$(ax + b)(cx + d) = acx^2 + adx + bcx + bd.$$

The two terms adx and bcx show that x is multiplied by ad and bc so these can be combined to read $(ad + bc)x$, so the final formula is:

$$(ax + b)(cx + d) = acx^2 + (ad + bc)x + bd.$$

There are no restrictions on the variables a, b, c and d. They can be any type of number, for example, positive or negative numbers, fractions or decimals. This means that only one formula is necessary and we just need to follow the normal rules of Arithmetic. A few examples should make this clear.

$$(3x + 2)(4x + 7) = 12x^2 + (3 \times 7 + 2 \times 4)x + 14$$
$$= 12x^2 + 29x + 14$$

$$(2x + 1)(3x - 5) = 6x^2 + (2 \times -5 + 1 \times 3)x + 1 \times -5$$
$$= 6x^2 - 7x - 5$$

$$(5x - 4)(2x - 3) = 10x^2 + (5 \times -3 + -4 \times 2)x + -4 \times -3$$
$$= 10x^2 - 23x + 12$$

It is worth mentioning at this stage that all the equations in this Chapter show answers found using Arithmetic even though x stands for an unknown number. This means that both sides of the equals sign must give the same answer for any value of x. If you are not clear about any of the results shown they can be checked by substituting a number of your choice for x.

In the last result we could perhaps choose to put $x = 8$. The left hand side would then be $(5 \times 8 - 4)(2 \times 8 - 3) = 36 \times 13 = 468$ and the right hand side becomes $10 \times 8^2 - 23 \times 8 + 12 = 640 - 184 + 12 = 468$.

One type of binomial multiplication is important enough to have special mention. This occurs whenever the two terms in one binomial are added and the same two terms are subtracted in the other. The simplest example of this type is:

$$(x + 1)(x - 1) = x^2 - x + x - 1$$
$$= x^2 - 1$$

The form of this result is always the same and can be easily expressed as a formula which includes all possible results.

$$(a + b)(a - b) = a^2 - ab + ba - b^2$$
$$= a^2 - b^2$$

It is well worth checking this result with some numbers:

$$(7 + 3)(7 - 3) = 7^2 - 7 \times 3 + 3 \times 7 - 3^2$$
$$= 49 - 21 + 21 - 9$$
$$= 40$$

If this is calculated by working out the contents of the brackets first we have:

$$10 \times 4 = 40.$$

This is an extremely important result, used widely in Arithmetic and Algebra, and is usually expressed the other way round as:

$$a^2 - b^2 = (a + b)(a - b)$$

Any expression of the type $a^2 - b^2$ is called, quite logically, a 'difference of two squares', and often arises in calculations involving right-angled triangles.

An example would be $28^2 - 26^2$. Without a calculator this would be a tiresome sum, but , using the above result it is only necessary to work out $(28 + 26)(28 - 26)$. This becomes $54 \times 2 = 108$. I will leave the reader to check that this is correct.

Division is best explained by a series of examples of slowly increasing difficulty. First, however, it would be useful to understand the structure of a division sum and note the correct names for each part of the sum. Division sums are set out exactly as in Arithmetic.

$$\text{Divisor} \,) \, \overline{\text{Dividend}}^{\,\text{Quotient}}$$

This can be written as an equation shown below.

$$\frac{\text{Dividend}}{\text{Divisor}} = \text{Quotient}$$

If both sides of this equation are multiplied by Divisor we have:

$$\frac{\text{Dividend}}{\text{Divisor}} \times \text{Divisor} = \text{Quotient} \times \text{Divisor}$$

After cancelling Divisor on the left hand side (as shown in *'Arithmetic'*) we have an important equation showing why the correctness of the answer to a division sum can be checked by multiplication.

$$\text{Dividend} = \text{Quotient} \times \text{Divisor}$$

Example 1

$$(4x + 6) \div 2$$

is set out as shown:

$$2 \,) \, \overline{4x + 6}$$

The first question to ask here is 'how many 2s do we need to make $4x$?', or put in another way 'what should 2 be multiplied by to make $4x$?'. Either way it should be clear that the answer is $2x$. This is now written in the answer space

$$2 \,) \, \overline{4x + 6}^{\,2x}$$

Attention must be paid to the signs when using Algebra and so the next question to be asked is 'what should 2 be multiplied by to make $+6$?'. The answer is $+3$ and so the sum is completed.

$$\frac{2x+3}{2\,)\,4x+6}$$

Example 2

$$2\,)\,\overline{6x-10}$$

If we multiply 2 by $3x$ the answer is $6x$, giving the first stage of the answer:

$$\frac{3x}{2\,)\,6x-10}$$

We now need to know what to multiply 2 by to get the answer -10. The correct number here is -5 (Multiplying and dividing negative numbers is fully explained in Chapter 6 of *'Arithmetic'.*) So the completed sum is:

$$\frac{3x-5}{2\,)\,6x-10}$$

Example 3

$$x+1\,)\,\overline{x^2+7x+6}$$

x has to be multiplied by x to make x^2, so $x+1$ must be multiplied by x and then subtracted as shown:

$$
\begin{array}{r}
x \\
\hline
x+1\,)\,x^2+7x+6 \\
x^2+x \\
\hline
6x+6
\end{array}
$$

Notice that the terms in x^2, x and the numbers are kept in their proper columns exactly as in Arithmetic. The sum is still not complete as we have a remainder which can be divided by $x+1$.

A further line is needed to show that $x+1$ has to be multiplied by 6 to obtain $6x+6$.

$$
\begin{array}{r}
x+6 \\
\hline
x+1\,)\,x^2+7x+6 \\
x^2+x \\
\hline
6x+6 \\
6x+6
\end{array}
$$

It is worth checking this answer by multiplication. If the formula is used we have:

$$(x + 6)(x + 1) = x^2 + (1 + 6)x + 6$$
$$= x^2 + 7x + 6$$

and so the answer is correct.

Example 4

$$x + 1 \overline{) x^2 + 7x + 3}$$

This one is to show how remainders can be dealt with. The first parts of the working are the same as the previous example and we can proceed to the last stage:

$$
\begin{array}{r}
x + 6 \\
x + 1 \overline{) x^2 + 7x + 3} \\
\underline{x^2 + x} \\
6x + 3 \\
\underline{6x + 6} \\
-3
\end{array}
$$

The answer can be left as $x + 6$, remainder -3, or the division by $x + 6$ can be indicated with a fraction, and so the complete answer can be written as:

$$x + 6 - \frac{3}{x + 1}$$

Example 5

$$x + 1 \overline{) x^2 - 1}$$

In this calculation $x + 1$ must be multiplied by x giving $x^2 + x$. However there is no x column in the dividend. In Arithmetic this column would be filled with a zero and the same can be done here by adding $0x$.

$$
\begin{array}{r}
x \\
x + 1 \overline{) x^2 + 0x - 1} \\
\underline{x^2 + x} \\
-x - 1
\end{array}
$$

Now x must be multiplied by -1 to obtain the final answer.

$$\begin{array}{r} x - 1 \\ x + 1)\overline{\smash{)}\,x^2 + 0x - 1} \\ \underline{x^2 + x} \\ -x - 1 \\ -x - 1 \end{array}$$

This sum has already been checked by multiplication on page 5.

Example 6

This final sum will be $(2x^4 + x^3 - 7x^2 - x + 5) \div (x^2 - 1)$. It will demonstrate a few points where care must be taken.

$$x^2 - 1)\,\overline{\smash{)}\,2x^4 + x^3 - 7x^2 - x + 5}$$

The first stage is to realize that x^2 must be multiplied by $2x^2$ in order to make $2x^4$ and so $(x^2 - 1) \times 2x^2 = 2x^4 - 2x^2$ is subtracted making sure that the terms are placed in the correct columns.

$$\begin{array}{l} 2x^2 \\ x^2 - 1)\,\overline{\smash{)}\,2x^4 + x^3 - 7x^2 - x + 5} \\ 2x^4 - 2x^2 \\ \hline + x^3 - 5x^2 - x + 5 \end{array}$$

$x^2 - 1$ must now be multiplied by x to make $x^3 - x$ which is subtracted.

$$\begin{array}{l} 2x^2 + x \\ x^2 - 1)\,\overline{\smash{)}\,2x^4 + x^3 - 7x^2 - x + 5} \\ 2x^4 - 2x^2 \\ \hline + x^3 - 5x^2 - x + 5 \\ + x^3 - x \\ \hline -5x^2 + 5 \end{array}$$

The final stage is to multiply $x^2 - 1$ by -5 giving $-5x^2 + 5$ thus completing the answer.

9

$$
\begin{array}{r}
2x^2 + x - 5 \\
\hline
x^2 - 1)\ \overline{2x^4 + x^3 - 7x^2 - x + 5} \\
2x^4 \qquad\quad - 2x^2 \\
\hline
+ x^3 - 5x^2 - x + 5 \\
+ x^3 \qquad\quad - x \\
\hline
-5x^2 \qquad +5 \\
-5x^2 \qquad +5
\end{array}
$$

It would be a useful exercise for the reader to check that this answer is correct by multiplying $2x^2 - x - 5$ by $x^2 - 1$. If you do not want to set out the formal calculation you can just multiply every term of $2x^2 - x - 5$ by x^2 and then by -1 giving six answers which can then be tidied up and put in the right order.

CHAPTER TWO

SOLVING LINEAR EQUATIONS

A linear equation is not very long.
There is one correct answer:
Any others are wrong.

It is first necessary to understand the general rules of equations. An equation consists of three parts:

1. Left hand side, usually abbreviated to LHS
2. Equals sign (=)
3. Right hand side, usually abbreviated to RHS.

It is important to realize that LHS must have exactly the same value as RHS in all the stages of solving an equation. At every stage in the process of solving an equation any change made to LHS must also be made to RHS. To demonstrate this with numbers seems rather pedantic, but it is worth stressing the obvious here.

If we start with $9 + 3 = 12$, this is clearly a true statement. If we add 5 to both sides we have $9 + 3 + 5 = 12 + 5$ and both sides are equal to 17. This equation is different from our original equation but it is still a true statement. Similarly, both sides of the original equation can be divided by 3 resulting in $3 + 1 = 4$. In general, if we perform the same operation to both sides of an equation we will get a new valid equation.

If we use this technique to solve the equation $2x - 3 = 5$, the first stage is to add 3 to both sides to get $2x = 8$. It is now a simple process to divide both sides by 2 showing that the solution is $x = 4$.

A linear equation is one that contains only numbers and terms involving x. The normal method of solution is to collect all the terms in x together to form one term and to combine all the numbers into one number.

Once this is done, the equation can be written as:

$$ax = b,$$

where a and b stand for any numbers (positive or negative, whole numbers or fractions).

Once the equation is in this form a formula can be obtained that will enable all linear equations to be solved.

Simply divide both sides of the equation by a, resulting in the formula:

$$x = b/a.$$

All that remains now is to show a few examples of linear equations and their solutions.

A fairly simple example would be:

$$8 + 5x = 3x + 12$$

If 8 and $3x$ are subtracted from both sides we get:

$$2x = 4.$$

Dividing both sides by 2 gives the answer $x = 2$.

In the next example some tidying is needed first:

$$5x - 9 + 7 - 7x = 14 - 4x$$
$$-2x - 2 = 14 - 4x$$

Adding $4x$ and 2 to both sides gives:

$$2x = 16.$$

Finally, dividing both sides by 2 shows that $x = 8$.

Care is needed when the equation contains fractions. In *'Arithmetic'* it was shown that if a fraction is multiplied by the denominator then the result is simply the numerator, e.g. $2/3 \times 3 = 2$. Equations involving fractions become easier to solve when the fractions are removed by multiplying both sides by the denominators. If there is more than one fraction it saves time if you can spot the lowest common denominator (see *'Arithmetic'*).

$$\frac{x}{3} + \frac{x}{6} = \frac{5}{2}$$

Multiplying both sides by 6 produces:

$$2x + x = 15$$

then
$$3x = 15$$

and the answer is:
$$x = 5$$

The unknown quantity, x, can also appear in the denominator as in the following example:

$$\frac{1}{x+1} + \frac{2}{x+1} = \frac{3}{5}$$

Multiply both sides by $x + 1$:

$$1 + 2 = \frac{3x + 3}{5}$$

Then multiply throughout by 5:

$$5 + 10 = 3x + 3$$

leading to $\qquad 15 = 3x + 3$

then $\qquad 12 = 3x$

and the answer $\qquad x = 4$

However, if the denominators involving x are different, another type of equation is revealed.

$$\frac{1}{x+1} + \frac{2}{x+2} = \frac{1}{5}$$

Start by multiplying both sides by $x + 1$ and by $x + 2$

$$x + 2 + 2x + 2 = \frac{x^2 + 3x + 2}{5}$$

Multiply both sides by 5:

$$5x + 10 + 10x + 10 = x^2 + 3x + 2$$

We now have a term involving x^2 which makes this a quadratic equation. The solution to this type of equation will appear in Chapter 4.

CHAPTER THREE

SQUARE ROOTS

Plus three is on the number line.
Minus three is also there.
To make everything fair,
They were twinned as a pair,
Known as the square roots of nine.

Whenever an operation (addition, multiplication, etc.) is defined in Arithmetic an opposite operation has to be defined in order to solve equations. This opposite operation is known as an inverse operation. You will already be familiar with addition and the inverse, subtraction; as well as multiplication and the inverse, division. In *'Arithmetic'* the operation of squaring (multiplying a number by itself) was explained. We will now deal with the inverse of squaring – finding a square root.

When trying to find the square root of a number we require another number which, when multiplied by itself produces the first number. Some square roots are quite easy to find. For example, the square root of 4 is 2, because $2 \times 2 = 4$.

There is a mathematical symbol to denote a square root. It is a tick ($\sqrt{}$), so we can write $\sqrt{4} = 2$.

Some square roots are fairly simple to spot, but others may require an informed guess and then have to be tested.

$\sqrt{9} = 3$ and $\sqrt{16} = 4$ are fairly easy, but $\sqrt{625} = 25$ may be a little more difficult.

Even with these simple square roots we have not yet obtained complete answers. If we think of $\sqrt{4}$, all we require is a number to be multiplied by itself to produce an answer of 4. It was explained in *'Arithmetic'* that if two negative numbers are multiplied then the answer is positive.

So, -2 multiplied by itself is also equal to 4. According to the definition the number 4 must have two square roots, +2 and -2. This will clearly apply to all positive numbers. This is where the plus or minus sign (±) is very useful.

We can now see the first few numbers that have whole number square roots.

$$\sqrt{1} = \pm 1$$
$$\sqrt{4} = \pm 2$$
$$\sqrt{9} = \pm 3$$
$$\sqrt{16} = \pm 4$$
$$\sqrt{25} = \pm 5$$
$$\sqrt{36} = \pm 6$$
$$\sqrt{49} = \pm 7$$
$$\sqrt{64} = \pm 8$$
$$\sqrt{81} = \pm 9$$

It is clear from the above list that the missing numbers (2, 3, 5, etc.) can not have square roots that are whole numbers. However, we can see that the square root of 2 must lie between 1 and 2 so it might be worthwhile checking if 1·5 is close to the right value. But 1·5 × 1·5 = 2·25, which is too big. It is logical to test 1·4. Trying this gives 1·4 × 1·4 = 1·96, which is too small.

So far we have found out that $\sqrt{2}$ must lie between 1·4 and 1·5. We could carry on with this process to get closer to $\sqrt{2}$ but it would soon become tedious. It is time to reach for the calculator. If I enter 2 on my calculator and press the square root button the answer is shown as 1·41421356237. This is an example of the calculator not being able to give a correct answer. The reason this is not correct will become clear if we try to test this answer by squaring. You will be relieved to know that only a small part of the multiplication need be attempted.

$$1 \cdot 4\,1\,4\,2\,1\,3\,5\,6\,2\,3\,7 \times$$
$$1 \cdot 4\,1\,4\,2\,1\,3\,5\,6\,2\,3\,7$$

$$\begin{array}{r} 9 \\ 0 \\ 0\,0 \\ \cdots\cdots\cdots\cdots \\ \cdots\cdots\cdots 9 \end{array}$$

We are only concerned with the right hand column of numbers. As shown in *Arithmetic* the 9 at the end of the third row above comes from 7 times 7 which is 49. All the other rows in the working have a zero in the right hand column. The 7 in the sum is in the

11th place of decimals, so without worrying about the other numbers our answer has a 9 in the 22nd place of decimals and so it cannot be exactly equal to 2, as hoped.

It does not matter where this trial and error stops. There will always be a number other than zero in the right hand column. No matter what single digit is squared the right hand column will contain either 1, 4, 5, 6 or 9. The only number that gives zero when squared is 0.

These paragraphs have demonstrated that $\sqrt{2}$ cannot be calculated exactly but they are not a proper mathematical proof. A proper proof will be shown at the end of this Chapter.

The same demonstration will also show that for all numbers that do not have a whole number square root, the exact square root cannot be calculated.

It seems that we have discovered a whole of set of numbers that cannot be calculated exactly. You may wonder whether these numbers actually exist. It is easy to show that they do by considering a square with side length equal to 2cm. This square will have an area of 4cm ^{2}. If you imagine this square gradually shrunk in size, while remaining a square, to have a side length of 1cm and area of 1cm ^{2}, then at some stage it must have an area of 2cm ^{2} and therefore a side length of $\sqrt{2}$cm.

Showing numbers on a number line was explained in *'Arithmetic'* and it was quite easy to show positive and negative numbers, vulgar fractions, decimal fractions (even recurring decimals) and percentage fractions. All of these numbers can be written as fractions (a/b). We now have these strange numbers which must be on the line but we do not know exactly where to put them.

It is necessary to find out how the operation of finding a square root fits in with the other operations of Arithmetic ($+$, $-$, $\times$, and $\div$). There are a few important things to remember that are best explained with simple examples.

$\sqrt{2} \times \sqrt{2} = 2$ (from the definition of square roots).

$\sqrt{4} \times \sqrt{9} = 2 \times 3 = 6 = \sqrt{36} = \sqrt{(4 \times 9)}$

In fact the last statement is true for all numbers, so it does not matter whether the multiplying is done before or after finding the square root. The proof of this is beyond the scope of this book, but you will find it to be true for all numbers that can be easily checked. It is also true for dividing.

Unfortunately, the same is not true when adding and subtracting are involved. Only one counter example is necessary to show this:

$\sqrt{4} + \sqrt{16} = 2 + 4 = 6$, but:

$\sqrt{(4 + 16)} = \sqrt{20}$, which is clearly not the same as 6 ($6^2 = 36$).

$\sqrt{20}$ can be made simpler by writing it as $\sqrt{(4 \times 5)} = \sqrt{4} \times \sqrt{5} = 2 \times \sqrt{5}$.

This last result is usually written as $2\sqrt{5}$.

Before starting on the promised proof that $\sqrt{2}$ cannot be calculated exactly it is worth a reminder of some facts about even numbers. These facts are so obvious that are in danger of being overlooked.

Firstly, if a number is even it can be divided exactly by 2 and can therefore be written as 2 times another number, e.g. $18 = 2 \times 9$.

Also, if the square of a number is even its square root must also be even. This is summarized in the table below.

$\times$	Odd	Even
Odd	Odd	Even
Even	Even	Even

This proof shows a powerful method whereby we assume that $\sqrt{2}$ can be calculated exactly and this then leads to a contradiction.

We assume that $\sqrt{2} = a/b$, where a/b is a fraction that cannot be simplified e.g. $3/2$.

If we square both sides we get $2 = a^2/b^2$ which can be written as $a^2 = 2 \times b^2$.

This means that a^2 is even and so a must be an even number. If a is even we must be able to write $a = 2 \times k$ (we do not need to know what value k has, merely that it exists).

In the equation $a^2 = 2 \times b^2$ we can replace a^2 by $4 \times k^2$ leading to:
$4 \times k^2 = 2 \times b^2$ and this simplifies to $2 \times k^2 = b^2$.

From what was shown previously about even numbers this means that b^2 is even and so b must be even. Now we have the contradiction. If a and b are both even then they can be divided by 2 to make the fraction a/b simpler, but we started by saying a/b was as simple as possible.

The only possible conclusion is that $\sqrt{2}$ cannot be written as a fraction and therefore cannot be calculated exactly

In addition to square roots we can have cube roots, fourth roots, etc.

For example, the cube root of 8 is 2 because $2 \times 2 \times 2 = 8$ and the fourth root of 81 is 3 because $3 \times 3 \times 3 \times 3 = 81$.

CHAPTER FOUR

SOLVING QUADRATIC EQUATIONS

A quadratic equation has terms in x squared.
There are always two answers, so be prepared.

The simplest type of quadratic equation contains only terms in x^2 and numbers and is fairly trivial. After any necessary tidying it would be possible to write an equation of this type as:

$$ax^2 = b$$

where a and b stand for any numbers.

This can readily be written in the form:

$$x^2 = a/b$$

and finally

$$x = \pm\sqrt{(a/b)}$$

Depending on the values of a and b we can either write down the answer or use a calculator.

Once we have terms in x as well as x^2 the solution requires special techniques. A fairly simple example is $2x^2 + 7x + 3 = 0$. No matter how we try to arrange this equation with the methods of Chapter 2 no solution is possible.

However, if we look back to page 3, we will see that :

$$2x^2 + 7x + 3 = (2x + 1)(x + 3).$$

The equation can then be written as:

$$(2x + 1)(x + 3) = 0.$$

This equation now is in the form of two numbers multiplied together giving an answer of zero.

Before attempting to solve the equation we need to look at the implications of two numbers multiplying to result in zero.

If we examine the equation:

$$5 \times b = 0$$

where b stands for any number we should see that the only way this can be true is if b $= 0$.

In a similar manner, the equation:

$$a \times 3 = 0$$

where b stands for any number, is only true when $a = 0$.

It is now only a small step to see that the equation:

$$a \times b = 0:$$

has two answers, namely $a = 0$ and $b = 0$.

We can now go back to the original equation:

$$2x^2 + 7x + 3 = 0,$$

which can be written as:

$$(2x + 1)(x + 3) = 0$$

This is the form of two numbers multiplied together giving an answer of zero.

It must, therefore, be true when:

$$2x + 1 = 0$$

and when

$$x + 3 = 0.$$

These two equations are linear equations and can be solved easily to give the two answers to the quadratic equation as $x = -1/2$ and $x = -3$. When these values are substituted in the original equation they prove to be correct.

The process of rewriting a quadratic equation as a product of two binomial expressions is called finding the factors, or factorising. It compares with finding the prime factors of numbers; explained in '*Arithmetic*'.

In order to find out how to do this we must look again at what happens to the numbers when two binomials are multiplied. In Chapter 1 it was shown that:

$(ax + b)(cx + d) = acx^2 + (ad + bc)x + bd,$

where a, b, c, and d represent any numbers, as usual. If we try to reverse this process it can be seen that the first numbers in each bracket (ax and cx) are the only possible factors of acx^2, and the second numbers (b and d) are the only possible factors of bd.

Clearly the process of finding the factors of a quadratic expression is not as easy as we might like.

As always in Mathematics if a problem proves to be too difficult we look at a simplified version in order to gain more of an insight.

The way this is done here is to try to factorise expressions where a and c are both equal to 1, thus making the problem much simpler.

The multiplication of the binomials now takes the form:
$$(x + b)(x + d) = x^2 + (d + b)x + bd,$$
Now to find the factors of $x^2 + 3x + 2$ we need to find two numbers that add up to 3 and multiply to make 2. A little thought should indicate that the numbers are 1 and 2. So the factors of $x^2 + 3x + 2$ are $x + 1$ and $x + 2$.

Multiplying $(x + 1)$ and $(x + 2)$ does indeed result in $x^2 + 3x + 2$. The equation
$$x^2 + 3x + 2 = 0$$
can be written as:
$$(x + 1)(x + 2) = 0.$$
This results in the two linear equations:

$x + 1 = 0$ and $x + 2 = 0$, and hence the two answers $x = -1$ and $x = -2$.
This becomes a little more difficult when negative numbers are involved. For example, $x^2 + 2x - 3$. We now require two numbers that multiply to make -3 and add up to 2. The factors of -3 could be -3 and 1 or 3 and -1. The pair that add up to 2 are 3 and -1, so the factors of $x^2 + 2x - 3$ are
$$x + 3 \text{ and } x - 1.$$
When a larger selection of factors is to be found there is a tidy way of setting out the calculations.

In the equation:
$$x^2 + 5x - 6 = 0$$
there are four pairs of factors of -6, namely, -6 and 1, 6 and -1, -3 and 2, and 3 and -2.

The following diagram shows the only possible factors of x^2 (x and x) on the top line. The first pair of factors of -6 (-6 and 1) are shown beneath the cross. The entries on the last line are obtained by multiplying along the arms of the cross to give $-6x$ ($-6 \times x$) and x ($1 \times x$) and then adding to get $-5x$.

Unfortunately, this is not the result we want. However, the diagram can easily be extended by adding further rows for the other three pairs of factors.

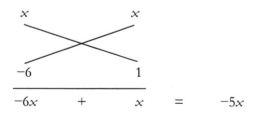

The completed table is shown below:

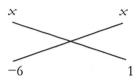

$-6x$	$+$	x	$=$	$-5x$
6		-1		
$6x$	$+$	$-x$	$=$	$5x$
-3		2		
$-3x$	$+$	$2x$	$=$	$-x$
3		-2		
$3x$	$+$	$-2x$	$=$	x

The results show the factors of four quadratic expressions, namely:

$$(x - 6)(x + 1) = x^2 - 5x - 6$$
$$(x + 6)(x - 1) = x^2 + 5x - 6$$
$$(x - 3)(x + 2) = x^2 - x - 6$$
$$(x + 3)(x - 2) = x^2 + x - 6$$

The original equation can now be written as:

$$(x + 6)(x - 1) = 0$$

with the two solutions $x = -6$ and $x = 1$.

This process has been a rather cumbersome way of solving the original equation and has also produced the surplus factors necessary to solve another three equations. It would be even more laborious if the first term had been $6x^2$, for example. We would then have to produce two diagrams for the pairs of factors of $6x^2$, namely $6x$ and x, and $3x$ and $2x$. Also the method would fail to find solutions for many equations, for example

$$x^2 + 4x + 5 = 0 \text{ (Try it and see!).}$$

The time has come to see if a general formula can be found for solving quadratic equations.

This formula can be found by closer examination of the multiplication when squaring a binomial. Some examples are shown:

$$(x + 3)^2 = x^2 + 6x + 9,$$
$$(x + 7)^2 = x^2 + 14x + 49,$$
$$(x - 2)^2 = x^2 - 4x + 16,$$
$$(x - 5)^2 = x^2 - 10x + 25.$$

Notice that in every answer, if we find the square root of the number at the end and double it, we get the number multiplying x. This confirms that we are quite justified in writing:

$$(x + a)^2 = x^2 + 2ax + a^2,$$

where a represents any number. This is known as a perfect square quadratic.

The next stage is to see whether the general quadratic equation, $ax^2 + bx + c = 0$ can be made to look a little more like a perfect square. We can certainly divide both sides by a, resulting in:

$$x^2 + \frac{bx}{a} + \frac{c}{a} = 0$$

To get the correct number term we need to divide the number multiplying x by 2 and then square it, i.e. $(b/2a)^2$ which can be written as $b^2/(4a^2)$. This term can be added to both sides of the equation to see if it helps

$$x^2 + \frac{bx}{a} + \frac{bx^2}{4a^2} + \frac{c}{a} = \frac{b^2}{4a^2}$$

If we now subtract c/a from both sides we will have a perfect square on the LHS.

$$x^2 + \frac{bx}{a} + \frac{b^2}{4a^2} = \frac{b^2}{4a^2} - \frac{c}{a}$$

This can be written as:

$$(x + b/2a)^2 = \frac{b^2}{4a^2} - \frac{c}{a}$$

The fraction c/a can be multiplied by 4a/4a to give equal denominators and the two fractions on the RHS can be combined:

$$(x + b/2a)^2 = \frac{b^2 - 4ac}{4a^2}$$

We can now find the square root of both sides of the equation. (Note that the square root of $4a^2$ is 2a).

$$(x + b/2a) = \frac{\pm\sqrt{(b^2 - 4ac)}}{2a}$$

Finally, b/2a can be subtracted from both sides, giving the formula for solving quadratic equations:

$$x = \frac{-b \pm \sqrt{(b^2 - 4ac)}}{2a}$$

If a quadratic equation has simple factors then it should be easier to use the factor method of solution. The formula can then be used if factors cannot be easily found.

Usually a calculator will be necessary when using the formula and the values will have to be entered very carefully. In order to make the task easier your calculator should have the following keys: $+/-$ (this changes the sign of the number entered), $\sqrt{}$ (this displays the square root of a number), x^2 (shows the square of a number), Min (this will place the displayed number in the memory) and MR (this displays the contents of the memory).

As an example the solution to: $2x^2 + 7x + 5 = 0$ will be calculated in easy stages. Here we have a = 2, b = 7 and c = 5, so the calculation is:

$$\frac{-7 \pm \sqrt{(7^2 - 4 \times 2 \times 5)}}{2 \times 2}$$

The most important part of this calculation is $\sqrt{(7^2 - 4 \times 2 \times 5)}$ as it is needed twice. This can be done first and stored in memory, so the first job is to make sure the calculator memory is empty. One way of doing this is press 0 followed by Min. The entry is now 7, x^2, $-$, 4, $\times$, 2, $\times$, 5, $=$, $\sqrt{}$. The answer 3 is displayed and pressing Min will put this into memory. The display can now be cleared and the new entry is 7 followed by $+/-$, showing -7. Now press $+$, MR then $=$, then $\div$, 2, $\div$, 2, $=$ showing the final answer of -1. The last sentence can now be repeated but entering $-$, MR, instead of $+$, MR. and the second answer of $-2 \cdot 5$ is displayed.

Some quadratic equations will produce answers with many more decimal places than you need. These answers can be rounded to the accuracy required as explained in *'Arithmetic'*.

There is one group of quadratic equations that will show an error on your calculator when you press the square root button.

A simple example is the equation $x^2 + x + 1 = 0$. In this case a, b and c are all equal to 1. This means that $b^2 - 4ac$ has the value -3, which is clearly a negative number and at this stage of Algebra it is not possible to find the square root of a negative number.

The explanation of solutions to this type of equation will appear in Chapter 12.

Finally, you may wonder if there is any practical application where an equation has two solutions. A simple case is when you throw a ball into the air. The formula linking height above ground, speed at which the ball is thrown and time is:

$$s = ut - 5t^2$$

where s stands for height in metres, u represents speed in metres per second and t is time in seconds. This equation can be written as:

$$5t^2 - ut + s = 0$$

Suppose the ball is thrown upwards at a speed of 50 metres per second. The equation then becomes:

$$5t^2 - 50t + s = 0$$

A question that could be asked is: after what time is the ball 45 metres above the ground?

The equation to solve is now:

$$5t^2 - 50t + 45 = 0$$

which simplifies to:

$$t^2 - 10t + 9 = 0.$$

Putting the numbers into the formula:

$$t = \frac{10 \pm \sqrt{(100 - 36)}}{2}$$

giving answers of $t = 1$ and $t = 9$. So the ball will be 45 metres above the ground after 1 second on the way up and again after 9 seconds on the way down.

If this exercise is repeated for a height of 125 metres you will find that there is only one answer, namely $t = 5$. Strictly speaking a mathematician would say there are two identical answers. The reason why the two answers are the same is that after 5 seconds the ball has reached the maximum height of 125 metres and starts to fall.

A further calculation for a height of 150 metres means solving the equation:

$$5t^2 - 50t + 150 = 0$$

reducing to:

$$t^2 - 10t + 30 = 0.$$

The calculation is then:

$$t = \frac{10 \pm \sqrt{(100 - 120)}}{2}$$

and this leads to trying to calculate the square root of a negative number. This indicates that the ball cannot reach a height of 150 metres.

CHAPTER FIVE

MORE ABOUT INDICES

An index is small and placed high on the right.
This does not reflect it's power and it's might.
It can make numbers large or make them small.
But on zero and one there is no change at all.

To open this Chapter we will start with a fairly simple idea and develop it to make some quite surprising conclusions. The following should be fairly obvious:

$$10^2 = 10 \times 10 = 100$$

We can now build on this with higher powers of 10:

$$10^2 = 10 \times 10 = 100$$
$$10^3 = 10 \times 10 \times 10 = 1000$$
$$10^4 = 10 \times 10 \times 10 \times 10 = 10000$$
$$10^5 = 10 \times 10 \times 10 \times 10 \times 10 = 100000$$

Over the centuries many important mathematical discoveries were made by examining patterns and there are some useful patterns in the above powers of 10.

Notice that in each row the index on the left is the same as the number of zeros on the right, e.g. $10^4 = 10000$.

Also to move down one row we must add one to the index on the left and add a zero (multiply by 10) on the right. Extending the rows in a downward direction following this rule can obviously be continued for as long as we like. Things become more interesting when we reverse this rule and move up one row – subtract one from the index on the left and divide by 10 on the right. The first time we do this is not very surprising:

$$10^1 = 10 = 10$$

To proceed further to the next row upwards we only need to show the left and right sides as the figure in the middle has no meaning.

When we subtract one from the index and divide by 10 to get the next row up the result appears rather strange:

$$10^0 = 1.$$

This result appears to be saying that if we multiply no 10s then the answer is 1. This is obviously something that we cannot do. The beauty of Mathematics becomes apparent here. If we have a new rule about numbers all that is required is that it is consistent with the established rules of Arithmetic and that it leads to something useful. Both of these conditions are satisfied here, so the above result is true without attempting to multiply.

Continuing the pattern upwards for one more line contains another surprise:

$$10^{-1} = 0 \cdot 1 = 1/10.$$

We now have a meaning for a negative index. The fraction at the end is included for a reason that will soon become clear.

It is now time to extend the rows sufficiently so that all the implications are clear.

$$10^{-5} = 0 \cdot 00001 = 1/100000$$
$$10^{-4} = 0 \cdot 0001 = 1/10000$$
$$10^{-3} = 0 \cdot 001 = 1/1000$$
$$10^{-2} = 0 \cdot 01 = 1/100$$
$$10^{-1} = 0 \cdot 1 = 1/10$$
$$10^0 = 1$$
$$10^1 = 10 = 10$$
$$10^2 = 10 \times 10 = 100$$
$$10^3 = 10 \times 10 \times 10 = 1000$$
$$10^4 = 10 \times 10 \times 10 \times 10 = 10000$$
$$10^5 = 10 \times 10 \times 10 \times 10 \times 10 = 100000$$

Again, any extension up or down is possible by following the two rules. It is still true that the index is the same as the number of noughts.

There is another pattern now appearing relating to negative indices, but first the use of a technical term makes the explanation simpler.

If we take a number and then divide 1 by that number the result is called the reciprocal of that number, so the reciprocal of 5 is 1/5. This is a mathematical operation and it was earlier stated that every operation must have an opposite. In this case it is it's own opposite, so the reciprocal of 5 is $1/5 = 0 \cdot 2$ and the reciprocal of $0 \cdot 2$ is 5. This also means that to find the reciprocal of a fraction you simply turn it upside down. (This may make you feel more comfortable with dividing fractions).

The pattern can now be described as the fact that a negative index is the reciprocal of the corresponding positive index, for example:

$$10^{-2} = 0 \cdot 01 = 1/100 = 1/10^2$$

We now come to the second requirement of a new rule – it must lead to something useful.

In many of the sciences extremely large and extremely small numbers have to be dealt with and to write these numbers in full would be rather tiresome. For example the mass of the Earth is 598000000000000000000000 kg (there should be 22 noughts here). Also the mass of an electron is 0·0000000000000000000000000000091 kg (31 noughts altogether).

Using powers of 10 these numbers can be written in scientific notation. The form this notation takes is a number between 1 and 10 (including 1 but not including 10) multiplied by a power of 10. A few examples follow:

$$100 = 1 \times 10^2$$
$$35000 = 3 \cdot 5 \times 10^4$$
$$0 \cdot 87 = 8 \cdot 7 \times 10^{-1}$$
$$0 \cdot 000025 = 2 \cdot 5 \times 10^{-5}$$

So the mass of the Earth can be written as $5 \cdot 98 \times 10^{24}$ kg and the mass of the electron as $9 \cdot 1 \times 10^{-31}$ kg.

If the process begun at the start of this Chapter were to be repeated with a number other than 10, say 7, we would establish a similar pattern where we would still add or subtract 1 to or from the index on the left, but multiply or divide by 7 on the right. A few rows are shown below:

$$7^{-3} = 1/343 = 1/7^3$$
$$7^{-2} = 1/49 = 1/7^2$$
$$7^{-1} = 1/7 = 1/7^1$$
$$7^0 = 1$$
$$7^1 = 7$$
$$7^2 = 7 \times 7 = 49$$
$$7^3 = 7 \times 7 \times 7 = 343.$$

The strange thing now is that we have 10^0 and 7^0 both equal to 1. If we had any other number as our starting point we would have the same result. In fact any number to the power of zero is equal to 1.

This is often stated as a $^0 = 1$, where a stands for any number (positive or negative, whole number or fraction).

The next important result obtained from powers of numbers comes when we examine what happens when these number are multiplied and divided.

As usual we will start with simple examples and build on them to work out the full implications.

If we wish to work out $9^3 \times 9^5$ we can do it the hard way by expanding the two numbers and writing the calculation as $(9 \times 9 \times 9) \times (9 \times 9 \times 9 \times 9 \times 9)$. Clearly, the brackets are not necessary and we just have eight 9s multiplied together which can be written as 9^8. This will still work if we change the 9 into another number and try different powers. The power, or index, just tells us how many of the numbers are being multiplied, so if we multiply by another power of the same number we can add the powers.

We now need to check whether this still works with negative indices, for example in the calculation $7^4 \times 7^{-2}$. We know that 7^{-2} is equal to $1/7^2$ so the calculation can be written as:

$$\frac{7 \times 7 \times 7 \times 7}{7 \times 7}$$

The two 7s in the denominator cancel with two of the 7s in the numerator leaving 7×7 which is the same as 7^2. So adding the indices gives the correct answer.

Dividing can be tested in the same way. $3^5 \div 3^2$ can be written as:

$$\frac{3 \times 3 \times 3 \times 3 \times 3}{3 \times 3}$$

As in the previous example, when the threes have been cancelled we are left with $3 \times 3 \times 3 = 3^3$; a result which could have been obtained by subtracting the indices. Using the idea of a reciprocal it is relatively easy to confirm that, for example, $4^5 \div 4^{-2} = 4^7$.

We have now worked out a rule for multiplying and dividing different powers of the same number, namely, when multiplying the numbers the indices must be added and when dividing the numbers the second index must be subtracted from the first.

One thing to be remembered is that in general the numbers must be the same. So this rule is of no help with a calculation like $5^4 \times 7^{-2}$. However, sometimes we can make the numbers the same. An example is $3^4 \times 9^3$. Because $9 = 3^2$ we can write $9^3 = 3^2 \times 3^2 \times 3^2 = 3^6$, so:

$$3^4 \times 9^3 = 3^4 \times 3^6 = 3^{10}$$

Now, if this rule is to be compatible with the rest of Mathematics it must apply to all indices. We have demonstrated that it applies to positive and negative indices so now we must investigate what is meant by a fractional index.

The easiest place to start is to try to find a meaning for $2^{1/2}$. The clue here is to use the fact that $1/2 + 1/2 = 1$.

Following this reasoning we can write:
$$2^{1/2} \times 2^{1/2} = 2^1 = 2.$$
This corresponds to the definition of the square root of 2. So $2^{1/2} = \sqrt{2}$.

Following this reasoning it can be shown that the index $1/3$ corresponds to the cube root. This can easily be checked with numbers that have whole number cube roots, for example, 8 or 27.

It is now only a small step to realise that an index of $2/3$ would mean the square of the cube root or the cube root of the square. The same answer is obtained whichever way the calculation is done.

The rules of indices can now be summarized as follows:
$$a^b \times a^c = a^{b+c}$$
$$a^b \div a^c = a^{b-c},$$
where a, b, and c represent any number.

For readers old enough to remember tables of logarithms, used as aids for multiplication and division, this Chapter has, in fact, been an extension of these tables. The word logarithm comes from the Greek *logos* meaning ratio and *arithmos* meaning number.

Tables of commonly used logarithms (logs for short) were based on powers of 10 and used the fact that any positive number could be written as a power of 10. Some simple examples are $10 = 10^1$ and $1000 = 10^3$. This means that the log of 10 is 1 and the log of 1000 is 3. To be able to deal with all numbers the tables of logs were used. To show how this worked with a fairly simple calculation we will try $2·6 \times 80$.

First, the tables would be used to find the log of $2·6$, which is $0·415$ and the log of 80, which is $1·9031$. As these are the powers of 10, these two numbers must be added for multiplication, resulting in $2·3181$. Now the anti-log of $2·3010$ must be found to give an answer of $208·0$.

This answer has worked out correctly, but this is not always the case. The usual tables were calculated to an accuracy of only four digits which meant that most answers were only accurate to three digits.

If you have a calculator with a button marked log you can try this yourself.

Logarithms are no longer used for calculating but they remain of great importance in more advanced Mathematics.

CHAPTER SIX

INTRODUCTION TO GRAPHS

To find your place along a line,
A number line, it works just fine.
To know where you are all over the page,
A second line is all the rage.

In order to mark a position in one dimension a number line is perfectly suitable and only one number is needed, hence the term one dimension. However, if a position anywhere on the page needs to be recorded, it can be done with two number lines. They are usually at right angles to one another and cross each other at zero. Two numbers are then required and consequently the page is a two dimensional space.

Those readers who are familiar with grid references on a map will already be used to this idea. In the Ordnance Survey maps of the United Kingdom the complete area is first divided into squares with sides of 100 km, with each of these squares designated by two letters. These squares are then subdivided into 1km squares. These squares are numbered from 00 to 99 across the bottom and up the side of the map. The numbers across the bottom are called Eastings and the numbers up the side are called Northings for fairly obvious reasons. To locate a particular 1 km square the Easting is written first followed by the Northing, for example, 8763. This locates a map position to within 1 km. If greater accuracy is required an estimate is made of the position in the square to the nearest tenth so that if this previous reference was in the middle of the square it would become 875635. Because the whole country occupies several 100 km squares the complete reference requires the two letters to identify the exact position, for example, ST875645. If you want to check your map reading skills this grid reference is in a small village in Wiltshire.

Graphs can be used to display a great variety of quantities. One popular graph, often shown in television news programmes shows the changes in the rate of inflation with time. Most of the time in Algebra we are concerned with two variables, x and y.

A sample pair of number lines is shown below. These lines normally cross at zero on both lines but this can be modified if necessary. When used for graphs it is normal to call these lines axes. The horizontal line is the *x*-axis and the vertical line is the *y*-axis.

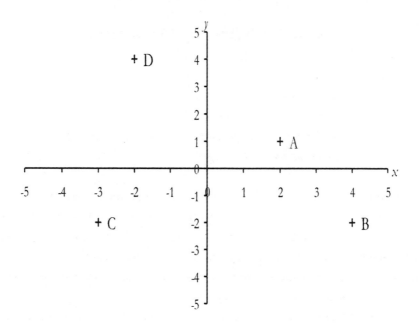

In order to locate positions of points within this framework two numbers are required. The number that is always given first is the distance along the *x*-axis and this is followed by the distance along the *y*-axis. These two numbers are enclosed in a pair of brackets and separated by a comma.

The positions of the points A, B, C, and D, above, are marked with crosses. A is the point (2, 1), B is the point (4, −2), C is the point (−3, −2) and D is the point (−2, 4).

The numbers in brackets are the Cartesian coordinates (just coordinates for short) of each point. As you may guess we have our old friend René Descartes (mentioned in the Introduction) to thank for the name.

Graphs of linear equations will be dealt with fully in the next Chapter, but it is appropriate to mention here some very simple graphs to illustrate how the equations are derived.

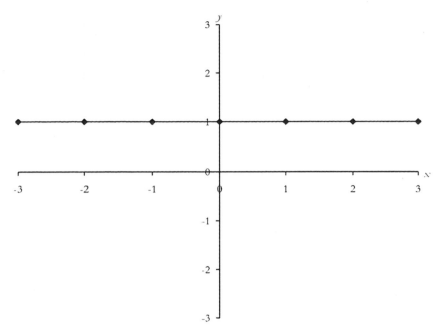

The coordinates of the points on the graph above are (−3,1), (−2,1), (−1,1), (0,1), (1,1), (2,1) and (3,1). The only thing that is constant here is that the y coordinate is equal to 1 in every case. It should be clear that any other point on the straight line shown through these points will also have a y coordinate equal to 1. This means that at every point on the line we can say that y equals 1. In proper algebra terms this means that the equation of the line is $y = 1$.

If the above line were to be moved up one unit then all the y coordinates would be equal to 2 and the equation of the line would be $y = 2$. In fact, any horizontal line on the graph has equation $y = c$, where c is a constant equal to the y coordinate of any point on the line.

This leads to an inescapable fact that sometimes causes confusion. The equation of the x-axis has to be $y = 0$.

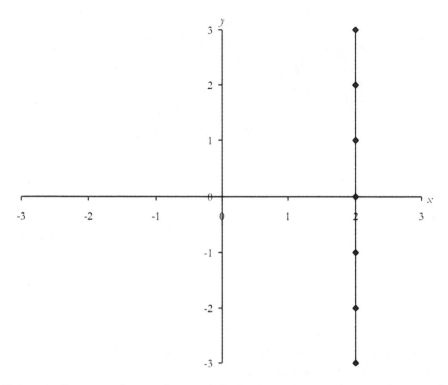

Using similar reasoning to that used for horizontal lines, the coordinates of the points marked on the graph above are (2,−3), (2,−2), (2,−1), (2,0), (2,1), (2,2) and (2,3). The only thing that is constant here is the x coordinate, which is equal to 2. Hence the equation of the line is $x = 2$.

If the line were to be moved three units to the left the x coordinates would all be −1, so the equation would be $x = -1$.

As before we must now come to the conclusion that any vertical line has the equation $x = c$, where c is a constant equal to the x coordinate of any point on the line.

Finally, this leads to the conclusion that the equation of the y-axis has to be $x = 0$. The last simple, but very important, line to be considered in this Chapter is where the x and y coordinates are equal to each other. This line is shown below.

The coordinates of the marked points are $(-3,-3)$, $(-2,-2)$, $(-1,-1)$, $(0,0)$, $(1,1)$, $(2,2)$ and $(3,3)$. All points on this line will have x and y coordinates that are equal to each other so the equation must be $y = x$.

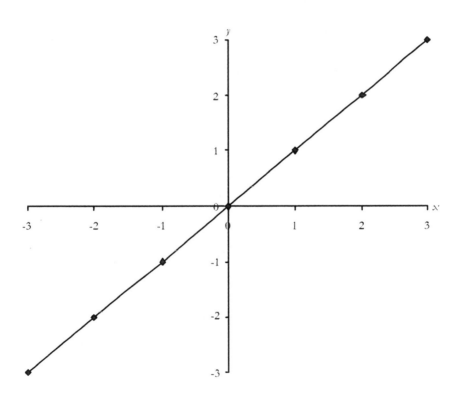

CHAPTER SEVEN

GRAPHS OF LINEAR EQUATIONS

A linear graph goes up hill or down dale,
But always it must be as straight as a rail.

The starting point for the examination of linear graphs is the graph of $y = x$, shown again below.

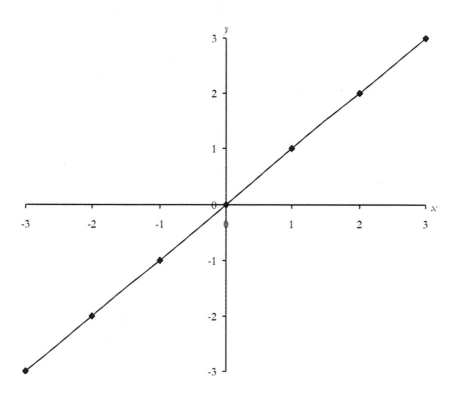

One of the properties that help to define this line is the steepness of the slope, normally called the gradient. Older readers may remember road signs warning of steep hills where the gradient would be given in the form '1 in 8'. This simply means that for every 8 units of distance travelled horizontally the road would rise by 1 unit. Modern signs now give the gradient as a percentage. In mathematics the gradient of a line is a number obtained by dividing the vertical distance by the horizontal distance.

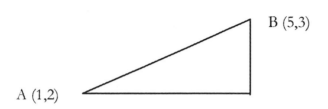

In the diagram above, when moving from point A to point B the vertical distance can be worked out from the difference in the y coordinates. This is equal to $3 - 2 = 1$. The horizontal distance comes from the difference in the x coordinates, $5 - 1 = 4$. So the gradient of the line from A to B is $1 \div 4 = 0 \cdot 25$.

One thing to notice here is that it does not matter which point is chosen first. If the order of the subtractions was reversed then the division would be $-1 \div (-4)$ which still gives an answer of $0 \cdot 25$.

The gradient of a straight line has to be constant at all points along its length so it does not matter which two points are chosen for the calculation.

With a little new notation a formula for the gradient of a line through two points can be derived. If the two points are called 1 and 2 the coordinates can be written as (x_1, y_1) for point 1 and (x_2, y_2) for point 2. The vertical distance between these two points is $y_2 - y_1$ and the horizontal distance between them is $x_2 - x_1$. The commonly used letter for gradient is m. So the gradient can be calculated from the coordinates using the formula:

$$m = \frac{y_2 - y_1}{x_2 - x_1}$$

It is worth noting at this stage that a positive answer for a gradient means that the line on the graph slopes up to the right whereas a negative answer shows a line that slopes down to the right.

If you check the gradient of the graph $y = x$ it is found to be 1. The next step is to examine a graph with a different gradient. The line $y = 3x$ is shown below:

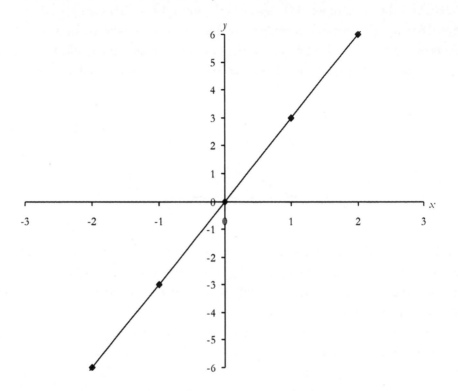

A check on the gradient between any two points on this line will give a result of 3.

From the two graphs shown so far in this Chapter you may well guess that the gradient of the line $y = mx$ is in fact m and you would be correct.

If we choose any two values for x, say, -3 and 2, the corresponding values of y are $-3m$ and $2m$, so using the formula for gradient we have:

$$\text{Gradient} = \frac{2m - (-3m)}{2 - (-3)} = \frac{5m}{5} = m$$

Changing the gradient of a line has the effect of rotating it with the origin as the centre of rotation.

If we now move a line up or down, the combination of these two changes will account for all possible straight line graphs.

Moving the line $y = x$ down 2 units gives the graph shown on the next page.

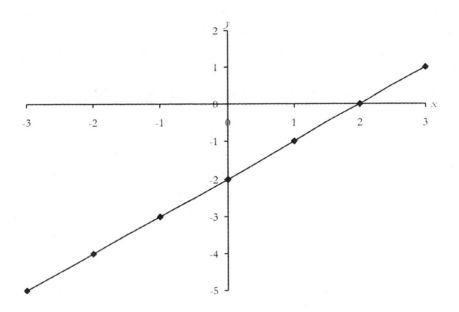

As all the y coordinates have had 2 subtracted, the equation is $y = x - 2$.

Moving the line $y = 2x$ up 1 unit produces the next graph.

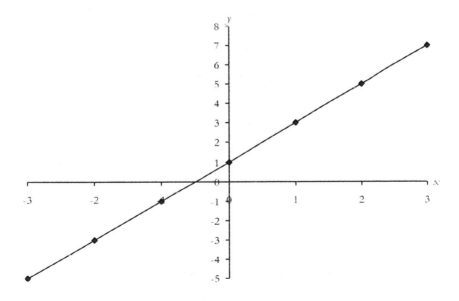

All linear graphs are completely defined when we know the gradient and the coordinates of one point on the line. It is normal to use the point where the line crosses the y-axis as the x coordinate at this point is 0. This point is normally called the intercept on the y-axis, or just intercept for short.

It is standard to use the letter c for the y coordinate of the intercept and so the equation of any linear graph can be written in the form:
$$y = mx + c,$$
where m and c can be any numbers, whole numbers or fractions, positive or negative.

A useful feature of graphs is that they can be used to solve a whole range of equations. Further examination of the first linear equation solved in Chapter 2, $2x - 3 = 5$, will demonstrate this. The expression $2x - 3$ is called a function of x. This simply means that the value of the function is only dependent on the value of x. When showing this on a graph we are setting this function equal to y and this enables corresponding pairs of x and y coordinates to be calculated.

The graph below shows $y = 2x - 3$ and $y = 5$ on the same set of axes.

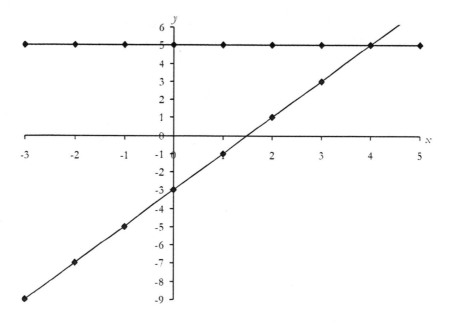

At the point where these lines cross the y coordinates are equal and so the x coordinate at this point (4) shows where $2x - 3$ is equal to 5 and the solution to the equation $2x - 3 = 5$ is $x = 4$.

This graph can also be used to solve $2x - 3 = 4$ or any other number on the RHS. In the real world, however, we would not bother to draw the line $y = 5$. It would be sufficient to find 5 on the y-axis, then move horizontally until the line $y = 2x - 3$ was reached and finally move vertically to the x-axis and record the corresponding x coordinate.

In the same way $2x - 3 = 4$ would be solved as $x = 3 \cdot 5$, whereas the equation $2x - 3 = -7$ yields the solution $x = -2$ and so on.

CHAPTER EIGHT

GRAPHS OF QUADRATIC EQUATIONS

I could not think of a decent rhyme.
I really did not have the time.
Suffice to say,
A quadratic graph is a curved line.
It will have a peak or a trough,
But only one at a time.

The simplest quadratic graph has the equation $y = x^2$, and is shown below.

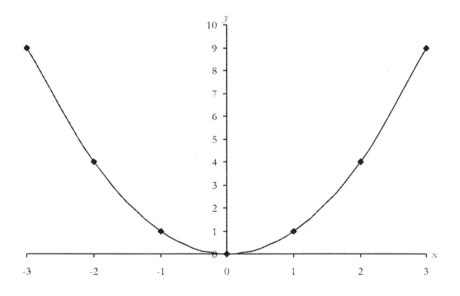

This graph has a trough. When the term in x^2 is negative a peak occurs as in $y = -x^2$, again shown on the next page.

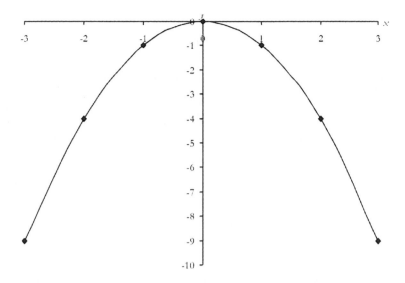

These shapes are the basis of all quadratic graphs. The general equation applicable to these graphs is $y = ax^2 + bx + c$, where, as usual, a, b, and c can be any numbers. As previously mentioned when a is positive the graph has a trough while a negative value results in a peak. The values of a, b and c will determine the steepness of the curve and the position of the curve relative to the axes.

The graph of $y = 2x^2 - 3x - 2$ is shown below to illustrate this.

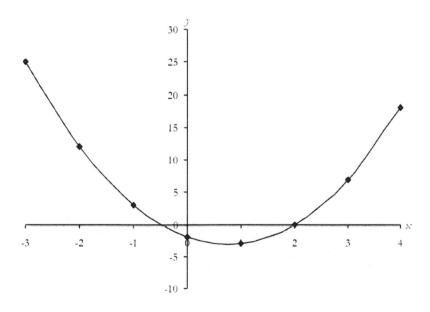

As mentioned earlier, a graph can be used to solve a group of related equations. The graph above can be used to solve equations where the function $2x^2 - 3x - 2$ is equal to any chosen number.

For example, to solve $2x^2 - 3x - 2 = 0$, the two solutions can be seen on the x-axis at $x = -0 \cdot 5$ and $x = 2$.

The solutions of $2x^2 - 3x - 2 = 12$ will be found roughly where $x = -2$ and where $x = 3 \cdot 5$. For accurate solutions finding factors or using the formula would still be necessary.

Even though the true quadratic function is $ax^2 + bx + c$, in order to get solutions to equations it is sufficient to use the graph of $y = ax^2 + bx$.

Using different letters (it is not compulsory to use x and y) and referring to the equation on page 24 about a ball being thrown up in the air, the graph needed is $s = -5t^2 + 50t$. This is shown below.

It is now easy to read off from the graph the height of the ball above the ground at

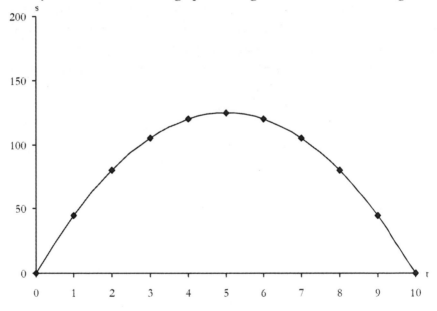

any time. The graph can also used to find the time for any particular height. The graph shows very well the fact that there are two times corresponding to any height and that the maximum height achieved is 125 metres.

If you need to draw graphs the first requirement is for some graph paper. This is sold in most stationers. The most common variety is likely to be ruled with 2mm, 10mm and 20mm squares and will look something like this:

The 10mm squares are slightly highlighted and the 20mm squares are even bolder for ease of reading scales on the axes. It is usually best to use the sides of the 20mm squares

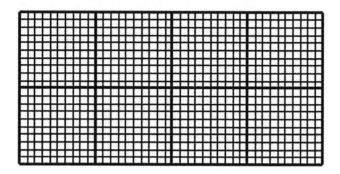

for the axes and also to mark the numbers at the 20mm squares. The 2mm squares then mark off tenths of the main unit.

The first task is to sort out a suitable scale so that the complete range of x and y values can be shown.

When drawing graphs of the form $y = ax + b$, the result is always a straight line so only three points need to be plotted, one at each extreme of the range of x values and one somewhere in the middle as a check on correctness. The points can then be joined with a ruler.

Graphs involving x^2, y^2 or higher powers require more care as these graphs will be curved lines, as will products of x and y. The number of points needed to plot graphs of the type varies and initially may be subject to trial and error. Extra points can always be plotted if necessary. Particular care is needed near turning points such as $(0,0)$ on the graph of $y = x^2$. When a graph changes direction it is always a smooth curve and not a sharp point.

It is possible to buy a set of curved templates called French curves to assist in drawing curved graphs. However they are not very easy to use. Experience and a steady hand will usually produce better results.

CHAPTER NINE

FINDING FACTORS

If a division sum works out exact,
You have found two factors and that's a fact.
It works with numbers and letters too,
With some practice 'you can do'.

Factorisation ranges from the very easy, e.g. $2x + 6 = 2(x + 3)$, to the seemingly impossible, e.g. $2x^4 + 19x^3 + 24x^2 - 73x + 28$. (The factors of this expression will be easily calculated at the end of this Chapter.)

Sometimes the terms of an expression may not be in the right order for easy factorisation and some rearrangement or simplification may be necessary. The expression $6 - 4x + 7x + 12$ can be written as $3x + 18 = 3(x + 6)$.

Sometimes two or more stages are needed to find all factors. The quadratic expression $x^2 + 5x - 6$ can be written as $x^2 + 6x - x - 6$. This now factorises as $x(x + 6) - 1(x + 6) = (x + 6)(x - 1)$. Spotting this type of factorisation needs experience and, fortunately, is not essential.

Finding factors of algebraic expressions is a powerful tool which often helps when trying to solve equations. To use this technique the RHS of the equation must be zero. We have previously seen that when the product of two expressions is equal to zero then solutions to the equation are found when either or both of the factors is zero. This was explained for two factors in Chapter 4 when solving quadratic equations.

This idea can be extended to as many factors as necessary. For example, if
$$a \times b \times c \times d = 0$$
Then the equation is satisfied when a = 0, b = 0, c = 0 or d = 0.

It is usual at this point to learn one more very useful piece of mathematical notation. It is a shorthand way of referring to an algebraic expression and it is written as $f(x)$ which means a 'function of x'. It is usually spoken as 'f of x'. Obviously if, for some reason, a different letter, say z is used then we would write $f(z)$.

If, for example, f(x) is defined as $2x^2 - 3x - 2$ we can then write the value of the function for different values of x in the following way:

$$f(-3) = 25$$
$$f(0) = -2$$
$$f(2) = 0$$
$$f(3) = 7$$

Notice that these values correspond to those that can be read from the graph of $y = 2x^2 - 3x - 2$ in the preceding Chapter.

It is now important to look again at the structure of a division sum. In Chapter 1 we started with:

$$\frac{\text{Quotient}}{\text{Divisor})\ \text{Dividend}}$$

This, we know can be rewritten as an equation:

$$\text{Dividend} = \text{Divisor} \times \text{Quotient}$$

At this point a new sign can be introduced. This is the identity sign ($\equiv$). The difference between this and the equals sign is that the equals sign is used when an equation has to be solved to find the correct solution or solutions, whereas the identity sign is used where the quantities either side of the sign are always equal no matter what value is given to x or other variable.

An example of where it could have been used earlier is:

$$(x + 1)(x - 1) \equiv x^2 - x + x - 1$$
$$\equiv x^2 - 1$$

The relationship between the parts of a division clearly must always be true so this can now be written as:

$$\text{Dividend} \equiv \text{Divisor} \times \text{Quotient}$$

However, this is not the complete picture as there is no guarantee that the division will be exact, so we must show the possible remainder.

$$\text{Dividend} \equiv \text{Divisor} \times \text{Quotient} + \text{Remainder}$$

We can use this to test whether $(x - a)$ is a factor of any algebraic expression, f(x), in the following way:

$$f(x) \equiv (x - a) \times \text{Quotient} + \text{Remainder}$$

Remember, we can choose any value for x here as we are dealing with an identity rather than an equation. If we choose the value a, we have a surprising result:

$$f(a) = (a - a) \times \text{Quotient} + \text{Remainder}$$
$$(a - a) = 0, \text{ and } 0 \times \text{Quotient} = 0, \text{ so:}$$

$$f(a) = \text{Remainder}$$

The result shows that if we substitute any number, a, for x in any expression then the answer is the remainder when the expression is divided by $(x - a)$. So if the answer happens to be 0 we have shown that $(x - a)$ is a factor of the expression.

This can first be tested with a simple example.

Suppose $f(x) \equiv x^2 - 3x + 2$. We can test any numbers we choose for a, but it is usual to try simple whole numbers first as this often gives clues if our first choices are incorrect. Some values of this function for small numbers follow:

$$f(0) = 2$$
$$f(1) = 0$$
$$f(2) = 0$$

We have been lucky here (mainly because this function was carefully chosen). The factors of $x^2 - 3x + 2$ are shown to be $(x - 1)$ and $(x - 2)$. I will leave it to the reader to check that $(x - 1)(x - 2) \equiv x^2 - 3x + 2$.

An important point can be demonstrated by finding the factors of the apparently similar function $f(x) \equiv x^2 + x - 2$ in the same way.

$$f(0) = -2$$
$$f(1) = 0$$
$$f(2) = 4$$
$$f(3) = 10$$

$f(1) = 0$ shows that we have one factor, $(x - 1)$, but as we increase the number tested we seem to be getting further away from zero. The answer is to try negative numbers. Trying -1 gives $f(-1) = -2$ and $f(-2) = 0$. This means that the second factor is $(x - (-2)) = (x + 2)$. Once again it is easy to test by multiplying that $(x - 1)(x + 2) = x^2 + x - 2$.

One more example to try before the 'seemingly impossible' one promised is $f(x) \equiv 2x^2 - 3x - 2$. Proceeding as before:

$$f(0) = -2$$
$$f(1) = -3$$
$$f(2) = 0$$
$$f(3) = 7$$
$$f(4) = 18$$

The answers are now getting further away from zero, so it is time to try negative numbers.

$$f(-1) = 3$$
$$f(-2) = 12$$

There appears to be something odd happening here as the answers are moving away from zero. If the positive and negative tests are combined in numerical order we have:

$$f(-2) = 12$$
$$f(-1) = 3$$
$$f(0) = -2$$
$$f(1) = -3$$
$$f(2) = 0$$
$$f(3) = 7$$
$$f(4) = 18$$

We have found one factor, $(x - 2)$, and it looks as though we should have another zero value between 0 and -1. The only sensible thing to do is to try $-\frac{1}{2}$. Sure enough $f(-\frac{1}{2}) = 0$, so the other factor is $(x + \frac{1}{2})$. However, this cannot be the complete answer as multiplication of these factors shows that $(x - 2)(x + \frac{1}{2}) = x^2 - 1\frac{1}{2}x - 1$. One of the factors must contain the term $2x$ as the only possible factors of $2x^2$ are $2x$ and x. We need to re-examine the equation

$$f(a) = (a - a) \times \text{Quotient} + \text{Remainder}$$

to see how this can be justified.

If we have correctly chosen a so that the Remainder is zero we can write

$$f(a) = (a - a) \times \text{Quotient} = 0$$

and it will also be correct to write

$$f(a) = 2(a - a) \times \text{Quotient} = 0.$$

Our second factor can now be written as $2(x + \frac{1}{2}) = (2x + 1)$.

The result of this factorisation is now: $(x - 2)(2x + 1) \equiv 2x^2 - 3x - 2$

At the start of this Chapter the reader was promised the easy factorisation of a seemingly impossible expression. So we start with:

$$f(x) \equiv 2x^4 + 19x^3 + 24x^2 - 73x + 28.$$

As before, we test for simple values of x that will give 0 as the answer.

$$f(0) = 28$$
$$f(1) = 0$$
$$f(2) = 162$$
$$f(3) = 700$$

We have found one factor, $(x - 1)$, but we now appear to be going in the wrong direction. It is time to try negative numbers.

$$f(-1) = 108$$
$$f(-2) = 150$$
$$f(-3) = 112$$
$$f(-4) = 0$$
$$f(-5) = -132$$

Another factor, $(x + 4)$, has been found. We clearly need more factors because when we multiply $(x - 1)$ by $(x + 4)$ the answer is $x^2 + 3x - 4$.

The answer will be found with a division sum as shown:

$$
\begin{array}{r}
2x^2 + 13x - 7 \\ \hline
x^2 + 3x - 4) \overline{\smash{)}\, 2x^4 + 19x^3 + 24x^2 - 73x + 28} \\
2x^4 + 6x^3 - 8x^2 \\ \hline
+ 13x^3 + 32x^2 - 73x + 28 \\
+ 13x^3 + 39x^2 - 52x \\ \hline
-7x^2 - 21x + 28 \\
-7x^2 - 21x + 28 \\ \hline
\end{array}
$$

The factors of $2x^2 + 13x - 7$ can be found to be $(2x - 1)$ and $(x + 7)$. We should now check the factors by multiplication. We must be certain that
$(x - 1)(x + 4)(2x - 1)(x + 7)$ is equal to $2x^4 + 19x^3 + 24x^2 - 73x + 28$

At least three multiplication sums are needed to check this result. Firstly $2x - 7$ can be multiplied by $x + 7$, giving an answer of $2x^2 + 13x - 7$. Then we can multiply $x - 1$ by $x + 4$ where the answer should be $x^2 + 3x - 4$. Finally we multiply $2x^2 + 13x - 7$ by $x^2 + 3x - 4$ and if everything goes to plan we have the final answer of $2x^4 + 19x^3 + 24x^2 - 73x + 28$.

For those readers who may wish to learn algebra beyond the level of this book or who may develop an interest in probability there is an alternative method relying on combinations.

In order to provide the answer to $(x - 1)(x + 4)(2x - 1)(x + 7)$ we must make sure we add all the possible combinations of multiplying single terms from each pair of brackets. It is best to first combine answers which result in the terms in x^4, then x^3, then x^2, then x and finally the numerical term. It is important to try to follow the pattern which builds up.

The term in x^4 can only be made by multiplying the term in x from each bracket, i.e. $x \times 2x \times x \times x = 2x^4$.

The term in x^3 can arise in four ways:
$$x \times x \times 2x \times 7 = 14x^3,$$
$$x \times x \times (-1) \times x = -x^3,$$
$$x \times 4 \times 2x \times x = 8x^3 \text{ and}$$
$$(-1) \times x \times 2x \times x = -2x^3, \text{ a total of } 19x^3.$$

The term in x^2 can arise in six ways:
$$x \times x \times (-1) \times 7 = -7x^2,$$
$$x \times 4 \times 2x \times 7 = 56x^2,$$
$$x \times 4 \times (-1) \times x = -4x^2,$$
$$(-1) \times x \times 2x \times 7 = -14x^2,$$
$$(-1) \times x \times (-1) \times x = x^2 \text{ and}$$
$$(-1) \times 4 \times 2x \times x = -8x^2, \text{ a total of } 24x^2.$$

The term in x can be found in four ways:
$$x \times 4 \times (-1) \times 7 = -28x,$$
$$(-1) \times x \times (-1) \times 7 = 7x,$$
$$(-1) \times 4 \times 2x \times 7 = -56x \text{ and}$$
$$(-1) \times 4 \times (-1) \times x = 4x, \text{ a total of } -73x.$$

Finally, the numerical term will be $(-1) \times 4 \times (-1) \times 7 = 28$.

Putting these five parts of the answer together we obtain:
$$2x^4 + 19x^3 + 24x^2 - 73x + 28,$$

which is the correct result.

The graph of this function is shown below showing clearly the values of x that give zero answers.

CHAPTER TEN

SIMULTANEOUS EQUATIONS

If x and y together appear
In an equation that's linear,
Many answers can be found.
Really this is not quite sound.
A second equation is needed that's clear,
To get a solution to appear.

When a situation arises in algebra where the values of two or more variables have to be found then a single equation is not enough. For example the equation $x + y = 10$ clearly is solved when $x = 2$ and $y = 8$, or when $x = 5$ and $y = 5$. In fact there are an infinite number of solutions. If we are also given the equation $x - y = 4$, you may already be able to see that the only values that satisfy both equations are $x = 7$ and $y = 3$.

One method for solving equations of this type is to rearrange one of the equations in the form $x = ?$ and then to substitute this value of x in the second equation. In order to keep track it is useful to number the equations. The example above can be solved in this way.

$$x + y = 10 \quad (1)$$
$$x - y = 4 \quad (2)$$

Equation (1) can be rewritten as $x = 10 - y$ and this in substituted in equation (2) to give $10 - y - y = 4$, leading to $y = 3$ and substituting this in equation (1) completes the solution, $x = 7$. These values for x and y are true at the same time for both equations, hence the name, simultaneous equations. This solution can be seen quite easily when both equations are rewritten in the form $y = -x + 10$ and $y = x - 4$ and shown on the same graph.

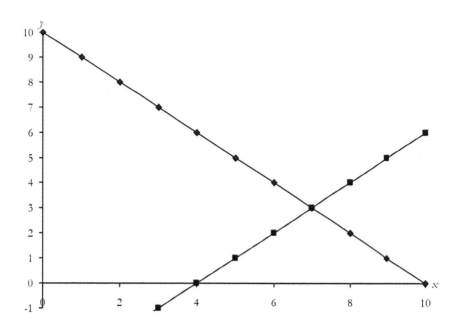

There is another way to solve these equations. Remembering that the equals sign means that the quantities on each side of the sign have the same value, we can add together equations (1) and (2) with the following result:

$$2x = 14$$

and so $x = 7$.

Subtracting equation (2) from equation (1) yields:

$$2y = 6$$

and so $y = 3$.

Sometimes, in order to use this method, one or both of the equations must be multiplied by a constant. For example:

$$2x + 3y = 7 \quad (1)$$
$$3x - y = 16 \quad (2)$$

One way of dealing with these equations is to multiply equation (2) by 3 and then adding the result to equation (1), producing the equation $11x = 55$, and so $x = 5$.

Substituting this value of x into equation (1) we have $10 + 3y = 7$.

This equation can be solved to give $y = -1$, thus completing the solution.

There is not always a solution to a pair of linear equations as is shown in the next example.

$$2x - y = 7 \quad (1)$$
$$4x - 2y = 8 \quad (2)$$

The reason that there is no solution becomes apparent when the two equations are shown on a graph:

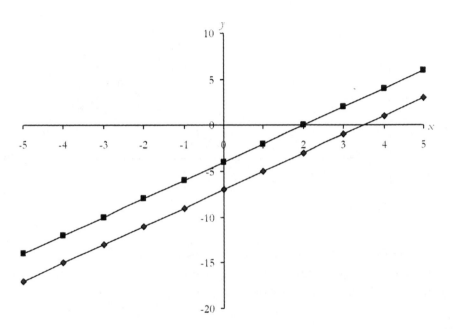

Here we have two parallel lines which, obviously, cannot cross each other and so there cannot be any pair of values of x and y on both lines.

It is now time to find a formula for solving a pair of linear simultaneous equations. In order to do this the equations need to be written in the form:

$$ax + by = c.$$

So we start with the equations:

$$ax + by = c \quad (1)$$
$$dx + ey = f \quad (2)$$

where, as usual a, b, c, d, e and f represent any numbers. If equation (1) is multiplied by e and equation (2) is multiplied by b we will be able to eliminate y by subtraction.

$$aex + bey = ce \quad (3)$$
$$bdx + bey = bf \quad (4)$$

Subtracting equation (4) from equation (3) results in:

$$aex - bdx = ce - bf \quad (5)$$

This can be factorised as:

$$(ae - bd)x = ce - bf \quad (6)$$

Dividing both sides by (ae − bd) gives the result for x.

$$x = \frac{ce - bf}{ae - bd}$$

To find the value of y we start again by multiplying equation (1) by d and equation (2) by a:

$$adx + bdy = cd \quad (7)$$
$$adx + aey = af \quad (8)$$

Subtracting equation (8) from equation (7) results in:

$$bdy - aey = cd - af \quad (5)$$

Once again, factorising produces:

$$(bd - ae)y = cd - af \quad (6)$$

Division then shows that:

$$y = \frac{cd - af}{bd - ae}$$

To make the format of this result similar to that of x it is convenient to multiply the RHS by -1/-1 (=1):

$$y = \frac{af - cd}{ae - bd}$$

If you test this result with any of the previous pairs of equations you will find that you obtain the correct solution except for:

$$2x - y = 7$$
$$4x - 2y = 8.$$

In this case ae − bd becomes $2 \times (-2) - (-1) \times 4 = -4 + 4 = 0$, and we cannot divide by 0. This confirms that the equations have no solution.

So far we have only dealt with pairs of linear equations. However, it is sometimes possible to find solutions to other pairs of simultaneous equations.

For example the following pair of equations can be solved.

$$x + y = 4 \quad (1)$$
$$\text{and } xy = 3 \quad (2)$$

From equation (1):

$$y = 4 - x$$

and this can be substituted into equation (2):

$$x(4 - x) = 3.$$

This equation can be rearranged to form the quadratic equation:

$$x^2 - 4x + 3 = 0$$

which has the solutions $x = 1$ and $x = 3$.

Substituting these values into equation (1) gives $y = 3$ when $x = 1$, and $y = 1$ when $x = 3$.

This interchangeability of results for x and y is because of the reversal properties of addition and multiplication. This was explained in 'Arithmetic'.

These solutions are shown below as the points of intersection of the graphs of $y = 4 - x$, (the straight line), and $y = 3/x$.

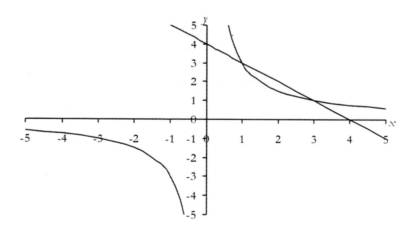

CHAPTER ELEVEN

IMAGINARY NUMBERS

x squared plus x plus one equals nought,
Has no solution, or so we are taught.
But the square root of minus one
Will ensure the job is properly done.

Before embarking on the explanation it will be useful to review the development of our number system so far.

In the dim and distant past the only numbers that had been invented were the positive whole numbers. As long as the only numerical task needed was counting in order to compare quantities, this set of numbers was quite sufficient.

Some time later it was thought necessary to be able to add these numbers and this formed the beginnings of Arithmetic. To make progress at this stage a symbol for zero had to be introduced and the idea of place value was helpful.

At a later stage when quantities needed to be shared the fractions had to be invented. To link this to solving equations in Arithmetic, this advance allowed equations like $3x = 4$ to be solved.

Soon came the invention of numbers smaller than zero, the negative numbers. This set of numbers was an introduction to the concept of abstract numbers. It is easy to see a positive quantity or a fraction of a quantity, but this is not possible with a negative quantity.

At this point Arithmetic was very useful and Algebra could be used to solve a huge variety of equations.

This was all very well until Arithmeticians realized that they could not solve the equation $x^2 = -1$.

It was decided to invent a symbol for $\sqrt{(-1)}$ and in Mathematics the symbol chosen is i (Engineers and Physicists often prefer the symbol j). This means that our set of available numbers has just been doubled. Every single number can be multiplied by i.

It is usual to write these numbers in the form $5i$, for example. These numbers are called Imaginary Numbers. Please do not be put of by the name. As mentioned earlier they are just as imaginary as negative numbers. They can be used in Arithmetic in the same way as any other number.

All that is needed to understand this new number is to realise that if it is to be of any use it must obey the rules of Arithmetic. The most important property of this number is revealed when we investigate powers of i.

Firstly, $i^2 = -1$ from the definition.

Next, $i^3 = i \times i^2 = i \times (-1) = -i$.

Finally, $i^4 = i^2 \times i^2 = (-1) \times (-1) = 1$.

If further powers are investigated, this cycle of results $(i, -1, -i, 1)$ is constantly repeated.

Because of this cycle of results simplifying the most awesome combinations of these numbers is extremely easy. The most complicated expressions can always be reduced to the form $a + bi$, where, as usual a and b stand for any number. Very often the first stage in simplifying expressions involving i is to replace all the powers of i by ± 1 or $\pm i$, as shown in the following example:

$$3i^5 - 2i^4 + 7i^3 + 8i^2 - 6i + 5 = 3i - 2 + 7(-i) + 8(-1) - 6i + 5$$
$$= 3i - 2 - 7i - 8 - 6i + 5$$
$$= -5 - 10i.$$

No other knowledge is needed in order to do addition and subtraction with imaginary numbers.

Multiplication is almost as simple. As any expression can be reduced to the form $a + bi$, we only need to consider multiplications of this type. A look back to Chapter 1 shows that we already have a rule for multiplying binomials and this must work here. If we multiply $(6 + i)$ and $(2 - 3i)$ we will obtain:

$$12 - 18i + 2i - 3i^2 = 12 - 18i + 2i - 3(-1)$$
$$= 12 - 18i + 2i + 3$$
$$= 15 - 16i.$$

It is fairly easy to obtain a formula for multiplication of imaginary numbers. We just multiply $(a + bi)$ and $(c + di)$, where a, b, c and d are our old friends, etc.

$$(a + bi)(c + di) = ac + adi + bci + bdi^2$$
$$= ac + adi + bci - bd$$
$$= (ac - bd) + (ad + bc)i.$$

Naturally enough, this Chapter is not all plain sailing. When it comes to division with imaginary numbers the usual division sum seems to lead nowhere.

This leads to a demonstration of one of the features of Mathematics, namely, if a problem appears too difficult then look for an easier way. The solution has already been seen in Chapter 1 and is a first class example of the use of the 'difference of two squares', To work out $(5 + 2i) \div (3 + i)$ we can write the sum as a fraction:

$$\frac{5 + 2i}{3 + i}$$

We know that multiplication by 1 does not change anything and the fraction:

$$\frac{3 - i}{3 - i}$$

is clearly equal to 1. If the two fractions above are multiplied, we can write:

$$\frac{5 + 2i}{3 + i} \quad \times \quad \frac{3 - i}{3 - i}$$

Using the rules for multiplying fraction this becomes:

$$\frac{(5 + 2i)(3 - i)}{(3 + i)(3 - i)}$$

The numerator becomes:
$$15 - 5i + 6i - 2i^2 = 15 + i + 2 = 17 + i,$$
While the denominator is equal to $9 - (i^2) = 9 + 1 = 10$.

The final result is then:

$$\frac{17 + i}{10}$$

By using the earlier information about the difference of two squares we have replaced an awkward division calculation with a much easier multiplication

For reassurance it is a good idea to check this result by multiplication. The necessary calculation is:

$$\frac{17 + i}{10} \quad \times \quad \frac{3 + i}{1}$$

Using the formula from the previous page this becomes:

$$\frac{(17 \times 3 - 1 \times 1) + (17 \times 1 + 1 \times 3)i}{10}$$

which simplifies to:

$$\frac{50 + 20i}{10} = 5 + 2i$$

and this is correct.

A formula for division with imaginary numbers can be constructed using the following sum:

$$\frac{a + bi}{c + di}$$

Using the difference of two squares method we have:

$$\frac{(a + bi)(c - di)}{(c + di)(c - di)}$$

Leading to the final answer:

$$\frac{(ac + bd)-(ad - bc)i}{c^2 + d^2}$$

We are now in a position to solve all quadratic equations. A reminder of the formula for solving quadratic equations allows the identification of three types of solution.

$$x = \frac{-b \pm \sqrt{(b^2 - 4ac)}}{2a}$$

The most important part of this formula is:
$$\sqrt{(b^2 - 4ac)}.$$
The explanation requires the introduction of two new mathematical signs. The first is >, which means 'greater than', as in $4 > 3$; and the second is <, which means 'less than', as in $3 < 4$.

When $b^2 > 4ac$, the expression is positive and has the normal positive and negative square roots.

When $b^2 = 4ac$, this part of the formula equals zero and so the quadratic equation will only have one answer ($-b/2a$). However, mathematicians insist that a quadratic equation must have two answers, and this situation is known as having two identical answers (also known as equal roots).

When $b^2 < 4ac$, the expression will be negative and i has to be used in finding the two square roots. The answers here will be of the form $a \pm bi$ and these numbers are called complex numbers and are explained in the next Chapter.

CHAPTER TWELVE

COMPLEX NUMBERS

Imaginary numbers and real,
Are put together with zeal.
Will we ever find an end,
To this very inventive trend?

It is useful to begin this Chapter with an update of the sets of numbers that have been defined so far, and to introduce some technical terms.

The simplest set of numbers consists of the numbers used for counting, 1, 2, 3, etc. This set of numbers is often referred to as the set of Natural numbers. The other numbers in common use are the sets of negative whole numbers and the set of fractions. The combination of these sets of number is called the set of rational numbers, and the test for a rational number is that it can be expressed in the form a/b, where a and b stand for whole numbers.

A set of numbers introduced in Chapter 3, the numbers we might well call the awkward square roots such as $\sqrt{3}$, is called the set of irrational numbers. This set also includes awkward cube roots, fourth roots, etc.

When the set of rational numbers is combined with the set of irrational numbers the result is called the set of real numbers.

The square root of minus one gives rise to the set of imaginary numbers, which contains rational or irrational numbers multiplied by the square root of minus one.

The set of numbers dealt with in this Chapter is formed by adding an imaginary number to a member of any of the previously mentioned sets of numbers. This is the set of complex numbers. As mentioned these numbers occur as solutions to quadratic equations when $b^2 < 4ac$ and examples of these equations follow.

The first equation to be solved will be the equation mentioned near the end of Chapter 4: $x^2 + x + 1 = 0$. In this case a=1, b=1 and c=1, so the answers are:

$$x = \frac{-1 \pm \sqrt{(1^2 - 4 \times 1 \times 1)}}{2 \times 1}$$

This simplifies to:

$$x = \frac{-1 \pm \sqrt{(-3)}}{2}$$

and finally:

$$x = \frac{-1}{2} \pm \frac{3i}{2}$$

Another example is: $x^2 - 4x + 13 = 0$. Here a = 1, b = −4 and c = 13. The solution is:

$$x = \frac{4 \pm \sqrt{(16 - 52)}}{2}$$

$$x = \frac{4 \pm \sqrt{(-36)}}{2}$$

Giving the final answers as $x = 2 \pm 3i$.

The final equation shows how the answers can be simplified. The technique relies on spotting a factor of ($b^2 - 4ac$) that has an exact square root. The equation is $3x^2 - 4x + 5 = 0$, so a = 3, b = −4 and c = 5. The stages of the calculation are:

$$x = \frac{4 \pm \sqrt{(16 - 60)}}{6}$$

16 − 60 equals −44, which is equal to 4 × 11 × (−1) so:

$$\sqrt{(-44)} = 2 \times \sqrt{(11)} \times i$$

The answers can then be simplified as:

$$x = \frac{2}{3} \pm \frac{2\sqrt{(11)}i}{3}$$

It should be noted here that the brackets indicate that the square root sign only applies to the number 11, and this would be calculated to any desired degree of accuracy.

The following graph showing the three types of quadratic equations referred to in

Chapter 11 (two real roots, two equal roots and two complex roots) illustrates the differences between them.

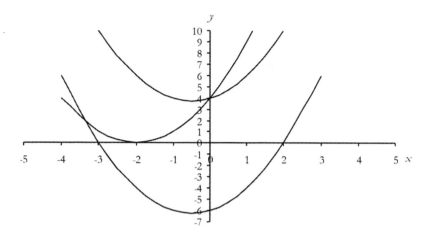

The lower graph is $y = x^2 + x - 6$ and the two roots of $x^2 + x - 6 = 0$ are clearly shown as $x = -3$ and $x = 2$.

The middle graph is $y = x^2 + 4x + 4$ and the equal roots of $x^2 + x + 4 = 0$ are both indicated at $x = -2$.

The upper graph is $y = x^2 + x + 4$ this does not cross the x axis and so has a pair of complex roots.

The number line has often been referred to in this book (and also in *Arithmetic*) and is a useful device for displaying numbers. Of course, imaginary and complex numbers have no place on this line. This difficulty is overcome by using a second number line for displaying imaginary numbers, having this line at right angles to the number line and arranging for the two lines to cross at zero on each line. The result is a graph with the number line for real numbers shown horizontally and the number line for imaginary numbers shown vertically.

The complex numbers $2 + 3i$, $-4 + i$, $-2 - 4i$ and $4 - 2i$ are shown on the graph on the next page.

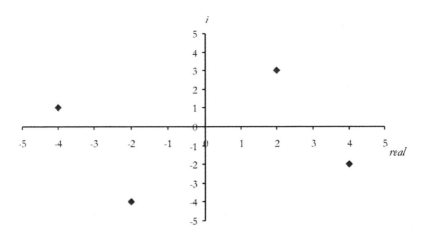

The invention of imaginary and complex numbers was originally to extend the boundaries of Mathematics. Some time later, however, it was seen to have practical uses. Without these numbers the development of modern electronics would not have been possible.

You may well think that we now have all the numbers that are necessary, but there are a few numbers that occur naturally that cannot be placed in any of the sets of existing numbers. These numbers are called transcendental numbers.

Most people have heard of one of these numbers. This is the number you get if you divide the circumference of a circle by it's diameter. No matter what size of circle is chosen the answer is always the same. Of course, if this number is worked out by measurement the answer can never be relied upon to be exact. The number is so important it has it's own symbol; the Greek letter π (pronounced pi). There are complicated ways of calculating this number and over the years groups of Mathematicians have used computers to get an ever increasing accuracy for the value of π. The present record seems to be about 10 trillion (1000000000000) decimal places, but this is constantly increasing. Most people are quite happy with the approximations of 3·14 or 22/7.

CHAPTER THIRTEEN

INEQUALITIES

The numbers 3, 2 and 1 were standing in a row.
3 turned to 2 and said 'I look down on you because you are smaller than me'.
2 turned to 3 and said 'I look up to you because you are bigger than me',
then turning to 1 'but I look down on you because you are smaller than me'.
1 turned to 2 and said 'I look up to you but not as much as I look up to him'.,
3 said 'I know my place'.

There is a lot of focus on equations in Algebra, but inequalities do have an important role as well. There are four types of inequality:

1. LHS larger than RHS, written as LHS > RHS,
2. LHS larger than or equal to RHS, written as LHS >= RHS,
3. LHS less than RHS, written as LHS < RHS and
4. LHS less or equal to RHS, written as LHS <= RHS.

When solving linear inequalities, the same number can added to or subtracted from both sides as in solving equations. Both sides can also be multiplied by or divided by the same positive number. However, when multiplying or dividing with a negative number the inequality sign has to be reversed. The reason for this becomes apparent by looking at the number line. As you move through zero on the number line the order of the numbers has to be reversed. One simple example should be sufficient to demonstrate this:

$$18 - 3x > 9$$

Subtract 18 from both sides: $\quad -3x > -9$

Divide both sides by -3: $x > 3$

This is clearly incorrect as a test using $x = 4$ will show. A reversal of the inequality sign gives the correct answer of $x < 3$.

Most of the time this problem can be avoided by doing the arithmetic in a different order, as shown below:

$$18 - 3x > 9$$

Add $3x$ to both sides: $\qquad$ $18 > 9 + 3x$

Subtract 9 from both sides: $\qquad$ $9 > 3x$

Divide both sides by 3: $\qquad$ $3 > x$

Which is clearly the same as: $\qquad$ $x < 3$.

Quadratic inequalities require a different technique to obtain a solution. One simple example is sufficient to demonstrate the method.

$$x^2 + x - 6 > 0.$$

The LHS can be factorised:

$$(x - 2)(x + 3) > 0.$$

If one of the pairs of brackets results in a positive answer and the other is negative, then the product is negative (< 0). So for this inequality to be true either the contents of both pairs of brackets must be positive, or they must both be negative. A little thought will lead you to the conclusion that either $x < -3$ or $x > 2$ are the solutions.

This result is shown clearly on the graph of $y = x^2 + x - 6$ below. The graph also shows the solution of $x^2 + x - 6 < 0$ as $x > -3$ or $x < 2$. This last result is usually combined in one statement as $-3 < x < 2$.

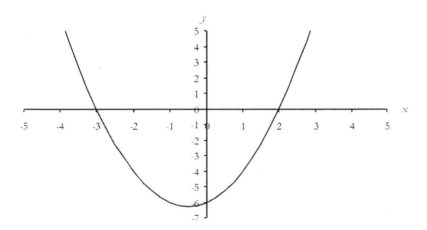

If the quadratic inequality had more difficult numbers so that only complicated factors could be found, then drawing the appropriate graph provides an easy method of solution.

Sometimes two unknown quantities have to satisfy several inequalities at the same time (simultaneous inequalities). A simplified practical example would be a company that operates a delivery service using two different sizes of van (called x and y). For a particular contract the company may need to know the different combination of vehicles they can use and which of these makes the greatest profit. The inequalities may well be:

$$x \geq 0$$
$$y \geq 0$$
$$x \leq 5$$
$$y \leq 4$$
$$2x + 5y \leq 16$$
$$3x + 2y \geq 5$$
$$x + y \leq 4$$

The first two inequalities show that only positive numbers need to be considered (fairly obvious as x and y refer to a number of vans). The second pair indicate that the company has 5 type x vans and 4 of type y. The fifth inequality could well have arisen from total journey times that should not be exceeded. The next could be because of the quantity of the delivery and the last is likely to be that there are only four drivers available. It looks as though a lot of thought may be needed here, but with the right lines on a graph it is very easy to see all solutions.

The x and y axes only need to show numbers from 0 to 5 and 0 to 4 in order to satisfy the first four inequalities. The line $2x + 5y = 16$ crosses the y axis at 3·2 and any solution must be on or below this line. $3x + 2y = 5$ crosses the y axis at 2·5 and solutions must be on or above the line. Finally, solutions must also be on or below the line $x + y = 4$.

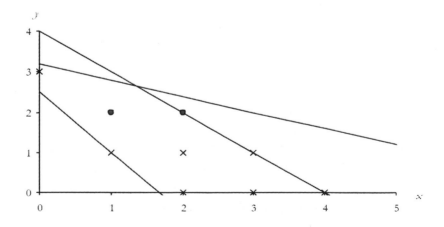

All possible solutions are defined by the five-sided shape with corners at (0,2·5), (0,3·2), (1·3,2·7), (4,0) and (1·7,0). All possible solutions have been marked on the graph and are summarized in the table below:

Type x vans	Type y vans
0	3
1	1
1	2
2	0
2	1
2	2
3	0
3	1
4	0

It can seen from the table that some solutions are superfluous. For example, the third entry suggests using 1 x van and 2 y vans, whereas the previous line shows that only 1 y van is needed.

It is a simple matter to delete unwanted solutions and to work out maximum profit.

Suppose the formula for calculating profit is $3x + 2y$. The final table below shows all the sensible solutions and the corresponding profit.

Type x vans	Type y vans	Profit
0	3	6
1	1	5
2	0	6

Clearly, the best solution is $x = 2$ and $y = 0$ as only 2 vans are used.

CHAPTER FOURTEEN

INTRODUCTION TO MATRIX ALGEBRA

A matrix is a rectangular array,
Of numbers or variables so they say.
Even though they are mathematical,
They do lead to things that are practical.

The main reason for including this Chapter is so that any readers who have hopefully developed an interest in the subject can see how using matrices (plural of matrix) with an addition to the rules of Arithmetic can lead to something new and also useful.

A matrix is defined as a rectangular array of numbers enclosed in brackets. Individual numbers in a matrix are referred to by the position they occupy, for example row 2, column 3. The abbreviations R2C3, etc. will be used from now on.

When referring to a particular type of matrix the description is the number of rows by the number of columns, so a 3 × 2 matrix has 3 rows and 2 columns.

The rules for adding and subtracting matrices is both simple and logical. The numbers in corresponding positions in the two matrices are added or subtracted to form a new matrix. For example:

$$\begin{pmatrix} 1 & 3 \\ -2 & 0 \\ 4 & -7 \end{pmatrix} + \begin{pmatrix} 3 & 4 \\ -1 & 2 \\ 5 & 6 \end{pmatrix} = \begin{pmatrix} 4 & 7 \\ -3 & 2 \\ 9 & -1 \end{pmatrix}$$

In order to follow this rule, the matrices to be added or subtracted must be of the same type. It is not possible to add a 3 × 2 matrix to a 2 × 2 matrix.

At this stage we need to look a little more closely at two more technical aspects of any type of arithmetic. In order to be consistent there must be a matrix which does nothing, known as the identity. For adding and subtracting in 'normal' arithmetic this is the number 0. With adding and subtracting matrices the identity is a matrix where all the numbers are 0.

The second requirement is that every matrix must have an inverse. The definition of the inverse is that when a matrix is combined with it's inverse the result is the identity. This sounds a little complicated but the idea is familiar in 'normal' arithmetic where the inverse of any number when adding is simply the negative of that number so the inverse of 3 is -3 and the inverse of $-1/2$ is $1/2$. The inverse of a matrix when adding follows the same pattern. To find this inverse all that is necessary is to replace every number with it's negative.

With subtraction in 'normal' arithmetic every number is it's own inverse, for example the inverse of -5 is -5 because $-5 - (-5) = 0$. Exactly the same is true when subtracting matrices.

The rule for multiplying matrices is rather different from 'normal' arithmetic for reasons which are beyond the scope of this book. It involves combining rows from the first matrix with columns from the second matrix. Matrix multiplication is only possible when number of columns in the first matrix is the same as the number of rows in the second matrix. The simplest possible example is to multiply a 1×2 matrix and a 2×1 matrix and this is shown below:

$$(1\ 2) \quad \begin{pmatrix} 3 \\ 4 \end{pmatrix} \quad = \quad \begin{aligned} &(1 \times 3 + 2 \times 4) \\ &= (11) \end{aligned}$$

In this case there is only one row in the first matrix and one column in the second so the explanation is reasonably clear. The first number in the row is multiplied by the first number in the column, the second number in the row is multiplied by the second number in the column, then the results of the two multiplications are added. It is essential to realise that matrix multiplication always means that rows from the first matrix are combined with columns of the second matrix. The rule can be expressed in symbols as:

$$(a\ b) \quad \begin{pmatrix} c \\ d \end{pmatrix} \quad = \quad (ac + bd)$$

Should the first matrix contain a second row the answer is shown below:

$$\begin{pmatrix} a\ b \\ e\ f \end{pmatrix} \quad \begin{pmatrix} c \\ d \end{pmatrix} \quad = \quad \begin{pmatrix} ac + bd \\ ec + fd \end{pmatrix}$$

Addition of another column to the second matrix gives the following result:

$$\begin{pmatrix} a\ b \\ e\ f \end{pmatrix} \quad \begin{pmatrix} c\ g \\ d\ h \end{pmatrix} \quad = \quad \begin{pmatrix} ac + bd & ag + bh \\ ec + fd & eg + fh \end{pmatrix}$$

The rest of this Chapter will concentrate on the square 2×2 matrix as this is one of the most important types.

The next essential is to find an identity matrix. In 'normal' multiplication this is the number 1, so it is worth trying the matrix where all the numbers are 1.

A simple test, using a general matrix with letters standing for any numbers shows that this, unfortunately, does not work:

$$\begin{pmatrix} a\ b \\ c\ d \end{pmatrix} \quad \begin{pmatrix} 1\ 1 \\ 1\ 1 \end{pmatrix} \quad = \quad \begin{pmatrix} a + b & a + b \\ c + d & c + d \end{pmatrix}$$

A close examination of this result reveals that replacing the 1 in the top right hand corner and the 1 in the bottom left hand corner with 0 is will give the correct result:

$$\begin{pmatrix} a\ b \\ c\ d \end{pmatrix} \quad \begin{pmatrix} 1\ 0 \\ 0\ 1 \end{pmatrix} \quad = \quad \begin{pmatrix} a\ b \\ c\ d \end{pmatrix}$$

The diagonal from top left to bottom right is important enough to have a name, the leading diagonal. In any square matrix the identity for multiplication has all the elements of the leading diagonal equal to 1 and all other elements equal to zero.

The best way to find out the rule for an inverse is to first show the link between square matrices and simultaneous equations.

The following pair of simultaneous equations were solved on page 53.

$$2x + 3y = 7 \quad (1)$$
$$3x - y = 16 \quad (2)$$

They are rewritten in matrix form below:

$$\begin{pmatrix} 2\ \ 3 \\ 3\ -1 \end{pmatrix} \quad \begin{pmatrix} x \\ y \end{pmatrix} \quad = \quad \begin{pmatrix} 7 \\ 16 \end{pmatrix}$$

If we knew the inverse of the first matrix in this equation we could multiply both sides by the inverse to get the answers for x and y. In the search for the rules for finding an inverse we will, as usual, start with a simple example

The inverse of $\begin{pmatrix} 2 & 1 \\ 1 & 1 \end{pmatrix}$ is assumed to be $\begin{pmatrix} a & b \\ c & d \end{pmatrix}$

where a, b, c and d have to be found.

This inverse must satisfy the matrix equation:

$$\begin{pmatrix} a & b \\ c & d \end{pmatrix} \begin{pmatrix} 2 & 1 \\ 1 & 1 \end{pmatrix} = \begin{pmatrix} 1 & 0 \\ 0 & 1 \end{pmatrix}$$

When the multiplication is done the equation becomes:

$$\begin{pmatrix} 2a + b & a + b \\ 2c + d & c + d \end{pmatrix} = \begin{pmatrix} 1 & 0 \\ 0 & 1 \end{pmatrix}$$

From this we have four equations available from which the inverse can be found:
2a + b = 1, a + b = 0, 2c + d = 0, and c + d = 1.

The first pair of equations give the result a = 1 and b = −1, while the second pair lead to c = −1 and d = 2.

So the inverse of $\begin{pmatrix} 2 & 1 \\ 1 & 1 \end{pmatrix}$ is $\begin{pmatrix} 1 & -1 \\ -1 & 2 \end{pmatrix}$

It seems as though the rule for finding an inverse may be as simple as interchanging the numbers on the leading diagonal and changing the signs on the other diagonal. However, as might be expected, one more test shows that this is not the complete picture.

$$\begin{pmatrix} 4 & 3 \\ 1 & 2 \end{pmatrix} \begin{pmatrix} 2 & -3 \\ -1 & 4 \end{pmatrix} = \begin{pmatrix} 5 & 0 \\ 0 & 5 \end{pmatrix}$$

The expected answer here is five times too big and the question now arises as to how the number 5 has appeared. An extremely important number associated with a square matrix is it's determinant. This number is calculated by multiplying the numbers on the leading diagonal and subtracting from this the product of the other two numbers. In the case the determinant is 4 × 2 − 3 × 1 = 5.

So, in addition to the number changes detailed above, an essential step is to divide by the determinant of the matrix.

In order to solve the simultaneous equations on the previous page we must first find the inverse of the matrix:

$$\begin{pmatrix} 2 & 3 \\ 3 & -1 \end{pmatrix}$$

The determinant here is $2 \times -1 - 3 \times 3 = -11$, so the inverse will be:

$$\frac{1}{-11} \begin{pmatrix} -1 & -3 \\ -3 & 2 \end{pmatrix}$$

Remember when a matrix is multiplied by it's inverse the result is the identity matrix which does nothing. So multiplying both sides of the matrix equation by the above inverse produces the following result:

$$\begin{pmatrix} x \\ y \end{pmatrix} = \frac{1}{-11} \begin{pmatrix} -1 & -3 \\ -3 & 2 \end{pmatrix} \begin{pmatrix} 7 \\ 16 \end{pmatrix}$$

so,

$$\begin{pmatrix} x \\ y \end{pmatrix} = \frac{1}{-11} \begin{pmatrix} -55 \\ 11 \end{pmatrix}$$

and,

$$\begin{pmatrix} x \\ y \end{pmatrix} = \begin{pmatrix} 5 \\ -1 \end{pmatrix}$$

leading to the correct answer of $x = 5$ and $y = -1$.

Before leaving matrix multiplication, it should be noted that, in general, the operation gives different answers depending on which matrix is placed first. The exceptions to this are when using the identity matrix and when multiplying a matrix and it's inverse.

Division of matrices is not defined as multiplication by an inverse is sufficient.

One application of matrix algebra has been demonstrated in this Chapter, but the topic has many more uses. An obvious extension to solving pairs of simultaneous equations with two variables is solving three simultaneous equations with three variables. For this the use of 3×3 matrices is required.

Another large topic where matrices are extremely useful is a branch of Geometry involving symmetry and transformations. 2×2 matrices are used for 2-dimensional Geometry and 3×3 matrices for 3-dimensional Geometry.

CHAPTER FIFTEEN

PARTIAL FRACTIONS

Splitting a fraction is the job to do.
The number of parts is at least two.
What we will do is have a guess.
If it doesn't work we're in a mess.
(Of course it will work!)

This Chapter is included for two reasons: splitting a fraction into parts that are added or subtracted leads to important results in more advanced Mathematics but, more importantly here is that it demonstrates the method of assuming the form of the answer and then testing for correctness.

Addition and subtraction of fractions was not dealt with in detail in Chapter 1 as the process is exactly the same as when using numbers. (See *'Arithmetic'* for extra help with this if needed.)

As a start we will see how a numerical fraction could be split up. The fraction 8/15 has a denominator with prime factors of 3 and 5, so it could have arisen from the addition of an unknown number of thirds and fifths. So we can write the sum:

$$\frac{8}{15} = \frac{A}{3} + \frac{B}{5}$$

and so:

$$\frac{8}{15} = \frac{5A + 3B}{15}$$

If two equal fractions have the same denominator, then the numerators must also be equal. This leads to the equation $8 = 5A + 3B$. It is fairly easy to see that A=1 and B = 1 is the simplest solution. There are other solutions, for example A = 4 and B = −4. We can now write the fraction 8/15 as:

$$\frac{1}{3} + \frac{1}{5} \quad \text{or} \quad \frac{4}{3} - \frac{4}{5}$$

Believe it or not the problem is easier to solve with algebraic fractions and only one answer is found. Just to check the process we will start with adding two algebraic fractions.

$$\frac{2}{x} + \frac{3}{x+1}$$

First the denominators are made the same:

$$\frac{2(x+1)}{x(x+1)} + \frac{3x}{x(x+1)}$$

Then, after adding and tidying, the final result is:

$$\frac{5x+2}{x(x+1)}$$

Looking at this process in reverse it should become apparent that if the denominator of a fraction has linear factors then it can be split into separate fractions with denominators being the linear factors. The problem is how to find the numerators.

This is where letters are used to stand for these numerators and then try to work out what they must be.

As we already know the answer it is sensible to start with the previous fraction and assume a result of the form:

$$\frac{5x+2}{x(x+1)} \equiv \frac{A}{x} + \frac{B}{x+1}$$

Note the use of the identity sign above which indicates that the equation must be true no matter what value is assigned to x.

The only progress we can make from here is to complete the addition and see if we have any clues as to the values of A and B.

$$\frac{5x+2}{x(x+1)} \equiv \frac{A(x+1)}{x(x+1)} + \frac{Bx}{x(x+1)}$$

Then:

$$\frac{5x + 2}{x(x + 1)} \equiv \frac{A(x + 1) + Bx}{x(x + 1)}$$

Now the denominators are the same so we can equate the numerators.

$$5x + 2 \equiv A(x + 1) + Bx.$$

If we now use the fact that we are dealing with an identity we can assign any value to x. Putting x equal to 0 shows that $A = 2$, and putting $x = -1$ gives the result $B = 3$, exactly as we hoped.

It is worth pointing out a short cut in the above process. As we know that the denominators on both sides are to be made equal, we can go directly from:

$$\frac{5x + 2}{x(x + 1)} \equiv \frac{A}{x} + \frac{B}{x + 1}$$

to:

$$5x + 2 \equiv A(x + 1) + Bx.$$

This is shown in the following example:

$$\frac{x + 2}{2x^2 + 5x + 3} \equiv \frac{A}{x + 1} + \frac{B}{2x + 3}$$

(To practice your skills you check that the denominator has been factorised correctly.)

By using the short cut we arrive at:

$$x + 2 \equiv A(2x + 3) + B(x + 1).$$

Using $x = -1$ leads to the solution $A = 1$ and putting $x = -3/2$ results in the answer $B = 1$, so the complete answer is:

$$\frac{x + 2}{2x^2 + 5x + 3} \equiv \frac{1}{x + 1} + \frac{1}{2x + 3}$$

It would be useful practice for the reader to check that this answer is correct.

The process of obtaining partial fractions can be extended to cover any number of linear factors. The next example shows how this works for three linear factors.

$$\frac{2x + 3}{x^3 + 2x^2 - x - 2}$$

Firstly, the denominator has to be factorised. Using the techniques explained in Chapter 9 these factors are found to be $x + 1$, $x - 1$ and $x - 2$.

So,

$$\frac{2x + 3}{x^3 + 2x^2 - x - 2} \equiv \frac{A}{x + 1} + \frac{B}{x - 1} + \frac{C}{x + 2}$$

The next step results in the identity:

$$2x + 3 \equiv A(x - 1)(x + 2) + B(x + 1)(x + 2) + C(x + 1)(x - 1).$$

Using the values $x = -1$, 1 and -2 in turn shows that $A = -1/2$, $B = 5/6$ and $C = -1/3$.

The final answer is therefore:

$$\frac{2x + 3}{x^3 + 2x^2 - x - 2} \equiv \frac{-1}{2(x + 1)} + \frac{5}{6(x - 1)} - \frac{1}{3(x + 2)}$$

The method explained in this Chapter will only work for linear factors. For quadratic factors and factors involving higher powers of x a modification is needed which is beyond the scope of this book.

A further proviso is that the numerator must be of a lower order than the denominator. If this is not the case then a division sum must be done and the remaining fraction is then split into partial fractions, for example:

$$\frac{4x^2 + 11x + 8}{2x^2 + 5x + 3}$$

In this case a division is possible and is shown below:

$$2x^2 + 5x + 3 \overline{\smash{\big)}\ 4x^2 + 11x + 8} \atop \ 2$$

$$\begin{array}{r} 2 \\ 2x^2 + 5x + 3 \overline{\smash{\big)}\ 4x^2 + 11x + 8} \\ 4x^2 + 10x + 6 \\ \hline x + 2 \end{array}$$

This gives an interim result of:

$$\frac{4x^2 + 11x + 8}{2x^2 + 5x + 3} \equiv 2 + \frac{x + 2}{2x^2 + 5x + 3}$$

But we already know that:

$$\frac{x + 2}{2x^2 + 5x + 3} \equiv \frac{1}{x + 1} - \frac{1}{2x + 3}$$

So, the final result is:

$$\frac{4x^2 + 11x + 8}{2x^2 + 5x + 3} \equiv 2 + \frac{1}{x + 1} - \frac{1}{2x + 3}$$

CHAPTER SIXTEEN

INFINITE SERIES

A division that goes on and on,
Does present some queries.
Where does it end? Nobody knows:
So the result is an infinite series.

If we look at divisions sums that have remainders in a different way – using negative powers of x – the division can be extended until a pattern emerges. Unusually for this book we start with a fairly difficult example so each step is explained in detail. The division sum shown on page 8 can be continued as explained below:

$$
\begin{array}{r}
x \;+\; 6 \\
x + 1 \;\overline{)\; x^2 \;+\; 7x \;+\; 3} \\
x^2 \;+\; x \\
\hline
6x \;+\; 3 \\
6x \;+\; 6 \\
\hline
-3
\end{array}
$$

The first question to ask is what should x be multiplied by to get the answer -3 and this is $-3x^{-1}$. So $x + 1$ is multiplied by $-3x^{-1}$ giving the result $-3 - 3x^{-1}$ which is then subtracted and $-3x^{-1}$ goes in the answer space.

$$
\begin{array}{r}
x \;+\; 6 \;-\; 3x^{-1} \\
x + 1 \;\overline{)\; x^2 \;+\; 7x \;+\; 3} \\
x^2 \;+\; x \\
\hline
6x \;+\; 3 \\
6x \;+\; 6 \\
\hline
-3 \\
-3 \;-\; 3x^{-1} \\
\hline
3x^{-1}
\end{array}
$$

We must now multiply $x + 1$ by $3x^{-2}$. This answer, $3x^{-1} + 3x^{-2}$, is subtracted and $3x^{-2}$ is added into the answer space as shown.

$$
\begin{array}{r}
x \quad + \quad 6 \quad - \quad 3x^{-1} \quad + \quad 3x^{-2} \\
\hline
x + 1 \,\overline{)\, x^2 \;+\; 7x \;+\; 3} \\
x^2 \;+\; x \\
\hline
6x \;+\; 3 \\
6x \;+\; 6 \\
\hline
-3 \\
-3 \;-\; 3x^{-1} \\
\hline
3x^{-1} \\
3x^{-1} \;+\; 3x^{-2} \\
\hline
-3x^{-2}
\end{array}
$$

Hopefully, the pattern is now apparent and we can give the answer as:
$$x + 6 - 3x^{-1} + 3x^{-2} - 3x^{-3} + 3x^{-4} - 3x^{-5} \ldots$$

This example was to show how division could be extended when there was no exact answer. An easier example now follows which does lead to a meaningful result. The sum is $1/(x-1)$.

$$x + 1 \,\overline{)\, 1}$$

Multiply by x^{-1}, subtract $1 - x^{-1}$ and place x^{-1} in the answer space.

$$
\begin{array}{r}
x^{-1} \\
\hline
x + 1 \,)\, 1 \\
1 - x^{-1} \\
\hline
x^{-1}
\end{array}
$$

Multiply by x^{-2}, subtract $x^{-1} - x^{-2}$ and place $+ x^{-2}$ in the answer space.

$$
\begin{array}{r}
x^{-1} + x^{-2} \\
\hline
x \;+\; 1 \,)\; 1 \\
1 \;-\; x^{-1} \\
\hline
x^{-1} \\
x^{-1} - x^{-2} \\
\hline
x^{-2}
\end{array}
$$

The sequence should now be clear and the result can be written as:

$$\frac{1}{x-1} \equiv x^{-1} + x^{-2} + x^{-3} \ldots$$

Note the use of the identity symbol here. As this is the result of a division sum it must be true for any value we wish to assign to x.

If we put $x = 10$ the resulting series may be familiar.

$$\frac{1}{9} = 10^{-1} + 10^{-2} + 10^{-3} \ldots$$

(back to the equals sign here).

This will be more recognizable when written as:

$$\frac{1}{9} = \frac{1}{10} + \frac{1}{100} + \frac{1}{1000} \quad \ldots$$

or in the form of a recurring decimal: (this topic was fully explained in 'Arithmetic')

$$\frac{1}{9} = 0 \cdot 111 \ldots$$

If we had started with $2/(x-1)$ then we would generate the sequence:

$$\frac{2}{9} = 0 \cdot 222 \ldots$$

and so on, up to:

$$\frac{9}{9} = 0 \cdot 999 \ldots$$

This last result is rather unusual as if we divide 9 by 9 the answer is clearly equal to 1.

Some people have a little difficulty with the concept of a recurring decimal being equal to a whole number. It is probably easier to accept by saying that the more 9s that are added to the recurring decimal the closer the answer approaches 1.

There is a similarity here to the children's conundrum which asks 'how long would a bar of chocolate last if you ate half of the existing amount every day?'.

In theory this requires finding solution for the sum of the infinite series:

$$\frac{1}{2} + \frac{1}{4} + \frac{1}{8} + \frac{1}{16} + \frac{1}{32} \cdots$$

It can be shown that the answer here is also 1. (Once again think of this as the answer gets closer and closer to 1 as the number of terms increase). Of course the implication is that the bar of chocolate would last for ever, but, in practice the portions would soon become too small to cut in half.

As usual there is some mathematical notation that makes writing the results of these series in a much shorter and more concise way. There is a symbol for infinity very much like a figure eight on it's side (∞). Also a horizontal arrow right facing arrow ($\rightarrow$) is used to mean 'approaches' or 'tends towards'.

If we use the letter S to represent the sum of the series:

$$\frac{1}{2} + \frac{1}{4} + \frac{1}{8} + \frac{1}{16} + \frac{1}{32} \cdots$$

and n for the number of terms used we can write:

As $n \rightarrow \infty$ S $\rightarrow$ 1. This would be read as 'As n approaches infinity S tends to 1'.

It was stated earlier that in the series for $1/(x - 1)$, x can be given any value we choose. However, if we put $x = 1$, we have the peculiar result:

$$\frac{1}{0} = 1 + 1 + 1 \ldots$$

The only sense we can make of this is that $\infty = \infty$, which is not very informative.

There are many infinite series that occur naturally in more advanced Mathematics. Special techniques have been devised in order to deal with these and many important results have been discovered.

I will finish with some closing words for the readers of this book and those who have also read 'Arithmetic'. The purpose of these books was an attempt to aid proper understanding of these two core subjects up to the standard of a good grade in a school leaving exam (GCSE at present). In 'Arithmetic', some topics were dealt with in a little more depth than required but in this book some parts were continued to a higher standard as it was a logical development to do so. Readers who go on to further study will probably appreciate this. For those satisfied with the school leaver's standard a knowledge of Geometry and Statistics with basic Probability is still required.